Praise for *Inmates in Charge*

It was a small town church in the Florida panhandle in 1998. White folks were making their way in for services when a distinguished-looking Black man and his wife walked in the door. There was a hush in the room. My first impression of Walter Beamon was, 'This guy is either very brave or he is lost.' I would soon learn that he was indeed very brave. So began a deep and abiding friendship that has spanned a quarter century. We have broken bread, laughed, cried, prayed, and fished. I love to hear him preach and sing. But most of all, I love his story. It is an American story that will stir your soul. With great admiration and respect, I call him my first cousin on the Holy Spirit side of the family. Listen to him.

Allen Bell,
Higher Ground, Florida

It was an honor to work for Chaplain Walt Beamon. As a young officer moving into leadership within the USAF Chaplain Corps, he taught me how to see with eyes focused on the potential of those I supervised . . . a quality he modeled in mentoring me. When I was right, he encouraged me. He was never afraid to help me with corrections when I was wrong. Fortunately, I call it blessed; I always received more encouragement than a correction from him. His focus was always on the purpose, the individual, and the process of doing right to get the best results. Walter Beamon cared for those he mentored because he was always focused on the future of the organization and the value of those that made up that organization . . . a wisdom that came from his experience. Though we shared differing racial and religious backgrounds, I am privileged and honored to call him "my mentor."

Richard F. Munsell,

Chaplain, Colonel, USAF (Retired)

Chaplain Walter Beamon had retired after a distinguished career in the U.S. Air Force Chaplain Corps. He was the highest-ranking active-duty Black colonel in the Air Force chaplaincy. When I met him, Walter and Ikie were attending the First United Methodist Church in Crestview, Florida, where I had recently been appointed Senior Pastor. Walter sang in the Chancel Choir, and they both taught Sunday School and Bible studies. Ikie was attractive, sophisticated, intellectually and socially, and a retired nurse. When I found out they had not been asked to unite with the church, we took care of that the following Sunday.

It could have been tense breaking the color barrier in those days and at that place, but it wasn't. We figured it was a "God-Thing!" Walter and I became good friends. We ate together in the local restaurants a couple of days a week. Our relationship was Christ-centered, warm, and fun. When we lost our young Assistant Minister, I wanted Walter to come on staff with me. Of course, we had to take the matter to the Administrative Board; we had already taken it to the Lord. Things were going to be tense, and we were a little nervous about how the Board would vote. Would we get enough votes? What a joy we felt when the Board's vote was unanimous in favor of Walter Beamon, another "God-Thing!"

It would have been a shame if Walter Beamon's talents had not been utilized. He was an excellent preacher, teacher, and pastor, and his pastoral prayers were moving and never repetitive. I am so glad he has written a book so that many can come to know this good man and what he has to say in Inmates in Charge. You will see that the author is motivated by a sense of fair play. I think this book is also a "God-Thing!"

The Reverend Dr. Richard S. Wright
Pastor and District Superintendent
United Methodist Church

In life, you are fortunate to find a true soulmate. Someone you know has their convictions and opinion on any topic long before you discuss it. Usually, it takes years together to develop such a relationship, not with Walter Beamon. I met him in my home church. Being a busy middle-aged Internal Medicine Physician, he is a retired Air Force Chaplain. It did not take long for me to know this man was a true brother in faith and conviction. He exudes joy, grace, and infectious peace, and knowing his background only confirmed to me his obvious call to ministry for all. I am glad he is writing this book because he will share the Gospel of grace and forgiveness that his life's work embodies as long as he has breath. As a White southerner, I will never know all his struggles, but I am a much better man and Christian because I encountered him. He is my soulmate, and we always pick up right where we left off with laughter, compassion, and understanding. Also of note, if you ever go fishing with Walter, do not let him call down the Holy Spirit to bless his bait; you will not catch nearly as many fish as he will.

Lee Thigpen, MD

But Jesus said to them, *"Watch yourselves carefully so you don't get contaminated with Pharisee yeast, Pharisee phoniness. You can't keep your true self hidden forever; before long you'll be exposed. You can't hide behind a religious mask forever; sooner or later the mask will slip and your true face will be known. You can't whisper one thing in private and preach the opposite in public; the day's coming when those whispers will be repeated all over town."*

Luke 12:2-3 (MSG)

I believe that Jesus is using Chaplain Beamon's extraordinary skills to expose the true heritage of the USAF Chaplaincy Service. ***Inmates in Charge*** brings the dirty little secrets of discriminatory practices into the cleansing light. Now, what was whispered in private will be shouted out in public. The book required a lot of hard work, courage, and perseverance. It is well written, researched, and documented. But its authenticity comes not just from what is written but from who wrote it. Chaplain Beamon is not only writing as historian but also an autobiographer. Many pages reveal not what he heard or researched but what he lived. This book is a treasured gift to the men and women who serve and who have served in the United Air Force Chaplaincy Service. It should be required reading for the Senior Leadership Course to the College of Chaplains.

Thank you Walt for your willingness to sacrifice so much to get this story told.

Chaplain, Lt. Col. (Ret.) Bennie R. Liggins

In this book, Walt tells his story and his personal experiences of inequities, double standards, and injustices in the USAF Chaplaincy. He risked his future with nothing to gain to become the voice and fighter for justice.

His impelling argument is statistical data that indicates how few African American Chaplains are promoted to Colonel when compared to overall promotions to Colonel in the USAF Chaplaincy.

In 1994, after 47 years, Walt was only the 9th African American Chaplain promoted to the rank of Colonel, and after 76 years, the USAF Chaplaincy has less than 30 African American Chaplains while hundreds and hundreds of Chaplains have been promoted to Colonel. There have been 25 Chaplains promoted to General Officer but not one African American Chaplain was deemed capable of that level of leadership.

He continually raised the red flag to leadership, who were also fellow clergy, but the issue was ignored or rationalized. Walt now takes up the mantel for equality and justice again for the last time, hoping someone in leadership will hear and take corrective action. Walt and I have been friends for over 30 years, and he is a man of God called to preach the Gospel.

John B. Ellington, Jr.
Chaplain, Major General, ANG, (Ret.)

Walt Beamon is a remarkable servant of the Lord Jesus the Christ. He reflects the fruit of the Holy Spirit in his life. I have known Walt for over 30 years. I am so glad he made the decision to write this book, **_Inmates in Charge_**. His career reflects the experience of most if not all African American Chaplains, including myself. I thank God for Walt for telling his story, and our story and the challenges we faced, in navigating through the system of systemic racism. I followed Walt in 1981 to Lakenhealth, England (Royal Air Force), and I encountered what he faced while there. I highly recommend this book for all to read. Walt lived the life as a Chaplain, worthy of his calling to serve the present age.

Chaplain Lieutenant Colonel,
Shelby B. Taylor, USAF (Retired)

Chaplain Colonel, USAF (Retired) Reverend Walter Beamon's book exposes a historical and insightful journey into the spiritual, controversial, and patriotic environment of African American Chaplains in the USAF. It promises to be a valuable reference tool. A must-read for any library. I'm eagerly awaiting its publication.

Chaplain Lieutenant Colonel
Theodore A. Henderson USAF (Retired)

INMATES
IN CHARGE

Top Level Leadership—
Lacking Vision, Corrupt, & Couldn't Be Trusted

Chaplain Colonel, Retired
WALTER BEAMON

KP PUBLISHING COMPANY

ISBN: 978-1-960001-45-0 (Hardcover)
ISBN: 978-1-960001-46-7 (Paperback)
ISBN: 978-1-960001-47-4 (eBook)

Library of Congress Control Number: 2024903403

Editor: Stacie Fujii
Cover Design: Juan Roberts, Creative Lunacy
Literary Director: Sandra Slayton James

Bible Permissions:
Scripture quotations marked NIV are taken from the Holy Bible, New International Version®, NIV®. Copyright © 1973, 1978, 1984, 2011 by Biblica, Inc.® Used by permission of Zondervan. All rights reserved worldwide.

Scripture quotations marked ESV are taken from the ESV® Bible (The Holy Bible, English Standard Version®). Copyright © 2001 by Crossway, a publishing ministry of Good News Publishers. Used by permission. All rights reserved.

Scripture quotations marked NKJV are taken from the New King James Version®. Copyright © 1982 by Thomas Nelson. Used by permission. All rights reserved.

Scripture quotations marked NLT are taken from the Holy Bible, New Living Translation. Copyright © 1996, 2004, 2015 by Tyndale House Foundation. Used by permission of Tyndale House Ministries, Carol Stream, Illinois 60188. All rights reserved.

Published by:

KP Publishing Company
Publisher of Fiction, Nonfiction & Children's Books
Las Vegas, NV 89117
www.kp-pub.com

Printed in the United States of America

CONTENTS

DEDICATION

This book is dedicated to the following people who have been pillars in my life. Whatever I have achieved, and whatever positive attributes I have made over the years, has been due to the impact of these individuals.

First, to my (deceased) dad and mother, Clement, and Earnestine Beamon, they were my nurturers. They were firm and loving parents who insisted that I learned the value of responsibility. They embraced core values that included truth, honesty, discipline, and hard work. They demonstrated Christ-centered lives for me to follow.

Second, Dr. Aavarah Strickland (deceased), my high school principal and a brilliant stalwart leader. He was instrumental in persuading my dad to allow me to attend Tougaloo College for my collegiate education. He instilled in me the hope and aspiration for a better life, which I needed at a pivotal time in my life.

Third, the Madison County Agriculture agent, Mr. Thomas Mackey (deceased). He provided an opportunity for me to grow wisely. He invited me to speak at a 4-H Club banquet, which proved to be the most embarrassing event in my life. However, that time was a blessing in disguise, which I will discuss in detail in this book.

Fourth, to my dear friend and comrade in the USAF Chaplaincy, Chaplain Colonel John R. Blair (Deceased). He was a strategic thinker, hard worker, and fought the good fight for justice and equality for African American chaplains.

Last, but certainly not least, this book is dedicated to the love of my life of more than sixty wonderful years, my soulmate and wife, Ikie. Over the years, I have recognized that she is smart, adventurous, and determined to achieve our goals. She has been a wonderful mother to our children, Angela (Angie) and Anthony (Tony), and a caring grandmother to our five grandchildren and three great-grandchildren. She is an educator, and a continues to be the "wind beneath my sails!" What more could one ask for?

FOREWORD

I received a call from my good friend, Chaplain, Colonel Walter (Walt) E. Beamon USAF, Retired, stating that he was going to write a book about his life while serving as a Chaplain in the United States Air Force Chaplains Corp. Now, his desire to write a book did not come as a complete surprise to me because previously, a group of us Black chaplains tried to write such a book, but our efforts failed. Walt, who was one of the architects of the project, was extremely disappointed. He often said to me, "It was a missed opportunity, but the world still needs to know our story."

I am Chaplain, Colonel Wilfred R. Bristol, USAF Retired. Walt and I served simultaneously in the Air Force Chaplaincy, at different bases, across the United States Air Force. Our paths crossed at the African American Chaplains Retreat, which the Air Force sponsored. The training was conducted by three giants in the Air Force Chaplaincy: Chaplain, Colonel, Raymond Hart (Deceased); Chaplain, Colonel, John Blair (Deceased); and Chaplain Beamon. The training and mentoring lasted for over 14 years; and provided African American Chaplains with the tools to compete with their peers for promotions.

The title of his book struck a chord with me, **Inmates in Charge**. This statement was frequently used by my mentor John Blair. I asked him, "What does it mean when you say that the "Inmates are in charge?" He explained that there is incompetence, discrimination and a lack of vision that exists at the highest levels of the Air Force Chaplains Corps. It is like the blind leading the blind. It is like the scripture says, "Where's there's no vision, the people perish." (Proverbs: 29:18)

Walt was at the tip of the spear and wore many battle scars fighting for equality as it related to Black chaplains getting the right assignments, sitting on Developmental Teams (DT), and receiving promotions to boards. It is important for you to know that if a chaplain does not get the right assignment, he or she will not go very far in their military career; and if a Black chaplain does not sit on your board, the greater the chance of not succeeding, because discrimination was one of the reasons why Black chaplains were failing. Due to his advocacy, Walt was threatened with letters of reprimand and counseling as a Colonel from his superior officers, but that did not deter him. Why, because he knew that the Air Force Chaplains Corps were consistently putting little people in big spots. Somebody once said, "I'd rather see a big person in a little spot, than a little person in a big spot."

Walt was a visionary and strategic leader that was respected by his peers and subordinates. He was often called to mentor Black, as well as, White chaplains. As he got ready to retire, the question arose as to who would train and mentor Black chaplains. Walt led the charge by organizing the United Military Chaplains Association. Through that organization, the training and mentoring continued for many years.

Walt's story is compelling. It is a must read. I dare say that the inmates are still in charge. In 2021 a legitimate (DT) board met to select chaplains for school assignments, which is prestigious if you are selected. They would also select alternates. If the primary could not go, the alternates would fill the slot. As a result of that board, the

2 alternate went to Air War College over the #1 alternate. There were no extenuating circumstances that prevented him from going to school. He was ready for duty. The inmates are still in charge.

Chaplain, Colonel Wilfred R. Bristol, USAF Retired

INTRODUCTION

Inmates In Charge: Top Level Leadership—Lacked Vision, Corrupt, & Couldn't be Trusted, a memoir of my life before, during and after my career in the United States Air Force Chaplain Service.

I became an active-duty chaplain in 1975, one of 26 African American chaplains throughout the entire Air Force since its founding in 1947. As you read this book, the title will take on a different perspective beyond the initial "prison" connotation. The book delves into a unique context, shedding light on the leadership dynamics within the highest echelons of the Air Force chaplaincy. We labeled those in leadership positions at the highest level of the Air Force chaplaincy inmates, and they were in charge. Why are they referred to as inmates? We called them inmates because they are the inheritors of a mentality which started before and during slavery and continues to this day. It is that mentality which is ingrained and exhibited through acts or behaviors by some Whites which clearly says to African Americans, "You are subservient to me, because I am White, and I am in charge!"

To illustrate this point, I will share an experience I remember as a child around 1958. My dad was driving to school in Camden, Mississippi, in our family's 1954 Chevrolet. While enroute to school, Dad had to pass the local cotton gin. An older White man was backing his truck onto the road. He backed into our vehicle and sustained a whiplash injury to his neck. He became irate and blamed Dad for the accident. He threatened to harm Dad and demanded that he pay him $25.00 monthly, indefinitely for his injury. Several times, when Dad passed the gin, he stood in the middle of the road with his shotgun in hand to intimidate Dad as he drove past him! The fact that he became irate, made threats, and demanded monthly payments, showed his imprisonment to a flawed reality. His behavior and emotional response demonstrated that he felt his White privilege and felt Dad was subservient to him because he was African American. He apparently believed that he could do anything he wanted, right or wrong, and nothing would be done about it! That was his reality. That is the mentality that imprisoned and still imprisons some Whites to this day.

Dad paid him for approximately three to four months and then decided he wouldn't pay him another dime. An African American man who worked for this White man contacted Dad and told him that this man was very angry and Dad needed to watch his back.

Dad bought a shotgun, a rifle, a 38 special handgun, and plenty of ammunition. Just in case, he had to defend himself and his family.

Jim Wallis, the well-known theologian, evangelical pastor, teacher, and fighter for social justice, wrote a book in 2016 entitled, *America's Original Sin-Racism, White Privilege, and the Bridge to a New America*. In his book on the subject of racism, I was struck by a statement that includes a question he made and it sheds light on the title and thrust of this book. He states in the book, *I'm Not a Racist, Am I?* This is a question that we hear from many White people. That is really the wrong question. It is not just an individual matter; these issues are historical and have to do with how our racial groupings as human beings have been deliberately manipulated for social and economic purposes.

Therefore, most of us have been socialized, instructed, and formed along the patterns of our racial groups, which is especially true for those of us who have been raised to think of ourselves as Whites."

In my memoir, *Inmates In Charge: Top Level Leadership—Lacked Vision, Corrupt and Couldn't Be Trusted,"* I recount my remarkable path from humble beginnings to breaking barriers in the military as a chaplain. Through my experiences, I shed light on the struggles, triumphs, and discrimination faced by African Americans in the United States Air Force.

My goals for writing this book are fourfold. First to share experiences and challenges of my military career with family, friends, and associates. Second, to expose the corruption, lack of vision, lack of integrity, and injustice of the chaplaincy leadership during my years of active duty. Third, to motivate Air Force Chaplain Leadership to devise a course of action that may ensure equal justice for all chaplains, and fourth, to demonstrate, through my life and testimony, that God's grace is sufficient and redemptive in the midst of all our struggles.

I hold the distinction of being the 9th African American chaplain in the United States Air Force's active duty history, to reach the rank of colonel, a milestone achieved by 28 individuals to date. Since my retirement more than twenty years ago, no African American chaplain on active duty has broken through the glass ceiling by becoming a general officer. What a travesty! More than twenty years ago, both the US Army and the US Navy promoted African Americans chaplains to the rank of major general.

I am sure that you, the reader, must be inquisitive as to why I am writing this book at this time in my life, at 83 years of age. To answer that question, I am completing a challenge that I established for myself on 10 December 2001 at my retirement dinner. I announced in my remarks, that one of my goals in retirement was to write my memoir.

During the first year following my retirement, I attempted to write a book. At that time, I felt emotionally drained, so I was incapable of writing this book. I will share more about that later. Instead of writing the book I focused on other ways that I could help active-duty chaplains enhance their careers.

In 2003, about a dozen active duty and retired African American chaplains met in Crestview, Florida to organize and form a non-profit 501(C) 3 organization.

That was the birth of the United Military Chaplains Association (UMCA). The primary purpose of UMCA was to be proactive in helping chaplains, especially African American chaplains, to realize productive and meaningful careers. Three of us senior ranking chaplains mentored them and helped them avoid some of the pitfalls we had experienced. We organized and held retreats at Hampton University in Hampton, Virginia, and other locations. We intentionally planned some of our retreats during the historical Hampton Minister's and Layman's Conference, which was held annually. This was done because the conference was successful and drew highly acclaimed African American preachers and theologians from throughout the country. One of our goals was to expose chaplains to the best preaching and theology possible. We believed it was an important experience for chaplains because it provided a venue for their spiritual support and encouragement. We also mentored them in areas that would provide them with some of the tools necessary to enhance upward mobility in their military career.

Another goal of the UMCA members was to write a book; one that would reflect our struggles and successes, our hopes and dreams. We wanted the book to serve as our legacy. However, our ghost writer could not complete the project, so the book was never written. Due to a lack of support from the Chief of Chaplains Office and from Command

Level, young African American Chaplains felt, by attending the retreats, that their careers were threatened. The pronouncement of the benediction of UMCA occurred in 2017.

In March 2022, Chaplain Colonel (Ret.) John Blair, a fellow Air Force African American chaplain and a dear friend, died. In early July 2022, I was enroute to his funeral at Arlington Cemetery. As I drove along the way, I reminisced about John and the days we struggled together as senior officers, trying to help those behind us—so many memories. Then I thought about the book we had hoped to write, and I silently said, "Well Lord, I guess it just wasn't meant to be," and moved on.

After John's burial, we held a Service of Remembrance in the Ft. Meyers Officers Club. Numerous chaplains and Line Officers, including General Officers, paid tribute to John's life and work. I was the last speaker on the program. I spoke of the challenges and struggles we had experienced as senior officers. John and I often saw things that were not right, especially at the Air Staff level and I would complain to John about it. His response, repeatedly were these words, "Walt, you know that the inmates are in charge!"

The question must be raised: what did John mean when he used that expression? I did not have to ask him what he meant. I knew him so well that there was no doubt in my mind what he meant. "Inmates in charge," in the context that he used it, referred to those in leadership positions at the highest levels of the Air Force chaplaincy, who were imprisoned by the historical flawed mentality of White privilege. They were self-serving, corrupt, lacked vision, and could not to be trusted. These people were power hungry and only cared for themselves and their cronies!

After the Memorial Service ended, I saw an African American woman, whom I had never seen or met before. She was an attendee. I had no idea what her connection with John might have been. She approached me, somewhat timidly and said, "Chaplain Beamon, while you were speaking, the Lord spoke to me and revealed that you are to write a book and the title is, "Inmates in Charge!" She further said, "I was hesitant to tell you, but I must be obedient to the Lord." My response was, "WOW" this is so interesting, because just yesterday, I was reminiscing about a book that our organization, UMCA had attempted to write and met with failure. I told the Lord I thought it wasn't meant to be."

When I realized what this lady said to me, it was like an affirmation that I should complete the long ago challenge I had set for myself in 2001 at my retirement dinner. I felt I had encountered an angel with a message for those few minutes. I said to her, "I would need a ghostwriter to help me write the book." She replied, "I work for a publishing company, and we have ghostwriters. When I get back home, I'm going to contact the publisher. You can expect a call from her within the next two weeks." The publisher called as promised. After submitting a bit of my work, she convinced me that I did not need a ghostwriter. This book is the fulfillment of my long-awaited dream. It is indeed a reality!

Chapter 1

EARLY YEARS

"Direct your children onto the right path,
and when they are older they will not leave it."

PROVERBS 22:6 (NLT)

I am the second son, but the only surviving son of my parents, Clement, and Earnestine Beamon. My older brother died at childbirth. I have four sisters: Myra, Agnes, Vivian, and Elaine. The first four of us are two years apart, until Elaine, who was born eleven years after Vivian.

I was born in Camden, Mississippi, a small rural community in Madison County on January 17, 1940. My birth certificate indicates that I was delivered by a mid-wife. Many, if not most African American women in the deep South used mid-wives during those times because they could not afford medical professionals and/or were not welcomed in hospitals. My mother told me that the day I was born was one of the coldest days she had ever experienced.

My sisters, Agnes, Elaine, Vivian, and Myra

My dad was the youngest in his family, so he inherited my grandparents' small home. It was an extremely modest home located near the end of a lane that branched off from the main county road. We had several black neighbors who lived past us on the lane, about a quarter of a mile away. The house was in the woods, surrounded mainly by evergreen and oak trees. There were wild fruit trees and wild berry bushes. We collected water for cooking and drinking from a freshwater spring a short distance from the house. When it rained, we collected water in large drums for bathing.

The foundation of the house was supported on concrete blocks. It was constructed of seasoned timber with a tin roof. The roof was rusted and often developed leaks requiring the use of pitched tar to prevent leaking. One of my favorite memories is of the times when I laid in bed listening to the rain drops falling on the tin roof.

The house had two bedrooms, which shared a double fireplace. The living room and kitchen were separated from the bedrooms by a hallway. The ceilings and walls of the bedrooms were adorned with flowered wallpaper. As a child I would visualize those flowers as I drifted off to sleep in my single bed. My sisters slept across the room on their larger double bed. As we got older my parents gave me more privacy by allowing me to sleep on the red plastic covered sofa in the living room.

My Dad, Clement Beamon

Electricity was not available in our community, so kerosene oil lamps provided the source of light for any indoor night chores and all homework during the school year. All outside chores had to be completed before I could begin homework. That meant I had to do my homework using the oil lamps. We had no indoor toilets. Our outhouse was the toilet.

Mama cooked all our meals on a cast iron cook stove. One of my chores was to cut and gather kindling and wood for the fireplace and the cook stove.

My earliest memories of my dad were of him farming our small farm and teaching elementary school. Most of the 80 acres were in pasture or timber land. We raised most of our food which consisted of corn, sweet potatoes, peas, butter beans, watermelons, sorghum cane, and all kinds of greens and other vegetables. We also raised a few chickens, hogs, and steers to provide meat.

Our financial crop was cotton. Our area was impoverished. A number of rural families were sharecroppers and worked in the fields of White farmers. They seldom made enough money to get ahead financially to own their own property. Many of my classmates could only go to school for four to five months during the school season, because they had to help on the owners' farms. This dynamic accounted for a high dropout rate for many Black junior and high school students. My siblings and I were blessed not to have to curtail our education because of farm work.

Prior to the early days of my elementary school, my dad received his certificate for teaching. To receive a teaching certificate, he had to finish high school and get his diploma. As far as I can recall, there were no high schools in rural Madison County for Black students during those days. Consequently, dad moved to New Orleans, Louisiana and lived with relatives to complete his high school education at McDonough

35 High School. This met the requirement for becoming a teacher. With a teaching certificate, teachers had to commit to continuing their education by attending college during their summers, until they graduated and received a bachelor degree. My dad received his degree in 1959, two years before I received mine in 1961.

My early years of school were especially memorable. Myra, my oldest sister, and I walked to school with our dad. Our school was a little two-room building on the grounds of our small Methodist church, Liberty Chapel A.M.E. Zion Church. Dad was the principal and there was one other teacher from the local community. After a couple of years, Dad was transferred to the little quaint town of Camden, Mississippi, located about four miles away. This time, our classes were held in the sanctuary of Murphy Chapel A.M.E. Zion Church. Again, Dad was the principal and Mrs. Lillian Conway was the other teacher. We were in the sanctuary because there were no other rooms except the hat and coat room and pastor's study.

There were probably 30 to 40 children of several grades in attendance. We sat on the pews in groups; one in the front of the sanctuary, the other in the back of the sanctuary. The year was around 1948. School in the sanctuary was a temporary situation because a new elementary school was in the works and would be located just across the road from Murphy Chapel. Our textbooks were "hand-me-downs" from the White Camden Elementary School. These books had been used for several years. Some were tattered, torn, ripped, and worn to shreds. The White children were bused to school in big yellow, comfortable buses. We had to walk. We had outhouses for our toilets, while White students had much better facilities. There were very few public schools for Blacks in Madison County. There were more in urban areas.

The typical day started with dad having to start a fire in the pot belly cast iron stove with coal. Then we would stand and recite the

Pledge of Allegiance and repeat the Lord's Prayer. Our teachers would discuss the lessons from the previous day, and ask for assigned homework. Then, some students would work on math problems while others practiced writing or reading with the teacher. We brought lunch from home in a small brown paper bag. Most of the lunches consisted of biscuits with sorghum syrup, a piece of pork sausage or strip of bacon, and a baked sweet potato. By my 10th birthday, or 5th grade our school had a concession stand which sold candy, salted peanuts, moon pies, nabs (crackers), and soft drinks. These items cost between a nickel or a dime. A nickel or dime was a lot of money for us during those days. We could only purchase from the concession stand periodically.

As I remember the rain falling on the tin roof, lulling me off to sleep, I also remember my dad's disciplining in the school. He set the rules and stuck by the rules for student inappropriate behaviors. Corporal punishment (whipping with belt or 5-foot saplings) was accepted for disciplinarian reasons during those days. Teachers and parents were on the same page when it came to discipline of children. Another method he used for discipline for boys only, was "head thumping" with his fingers. Believe me, you would not want dad to thump your head!

One of the school's rules was NO SMOKING! Sometimes during lunch hour, the boys would slip into the woods behind the church, find wild rabbit tobacco plants, strip off dry leaves, roll it up in paper and smoke it. I would often accompany them, but never tried to smoke it. One day they convinced me to try it. They set me up! As soon as I took one puff one of the boys ran back to church/school and told my dad that I too had smoked. That day would be a day that I would long remember.

My dad made an example out of me by whipping me before all my classmates. That day I felt my dad was more than cruel to me! He proved there was equal justice in the school. I learned my lessons the hard way by violating his rules. Even as an adult, I could not bring myself to smoke in his presence.

Our transportation at that time was very limited. In fact, my dad bought an old army jeep/truck with no doors.

He built an enclosure similar to a large box on the bed of the jeep to protect us from the elements during inclement weather.

Sometimes he picked other children up on our route to school. It was not heated, so on cold days it was still better than walking. Some days when the old jeep was not running, we had to walk. Sometime on rainy days when the school bus for White children passed us, it inadvertently splashed muddy water on us. Those were rough and difficult days.

During the late summer and fall, my daily routine went something like this: on the weekdays, Monday through Friday, I got up at about 5:30 a.m. I had to be an early riser because I had to milk five to seven cows, get back home, get dressed for school, eat breakfast, and walk four miles to school when the jeep was not running. School dismissed around 2:30 p.m. By the time I walked home, it was around 3:45 p.m. I would grab a snack such as a baked sweet potato, tea cake, or a left-over biscuit. Then I would take off my school clothes, put on my work clothes, and head to the field to pick cotton, harvest corn or other produce. When I left the fields I would go to the pasture, and steer the milk cows into the pen near the barn, I rushed to get home before dark.

One reason for rushing was because I was eager to eat Mama's supper. Those were the best meals, and I would be plenty hungry! Supper included fresh hot cornbread, butter, and buttermilk, and bacon or country ham with sorghum molasses. Often we had baked sweet potatoes from our farm. Man was that some good eating! After supper we would do our homework and go to bed. I often went to bed happy and satisfied, but very tired.

When I was about 12 years old, dad would travel back and forth during June and July working on his college degree. He took courses at Jackson State, Tougaloo, and Rust College. When he attended Jackson State and Tougaloo, he was home in the evenings. However, when he attended Rust College, he was away Monday thru Friday, because it was 150 miles away from home. He made me responsible for most of the work on the farm. This included plowing the crops using Ole Ben and Ole George, our mules. The sun was blazing hot.

Me at 12 years old

Sometimes, the temperature would reach 98 to 100 degrees, with humidity up to 95% or more. It seemed as though there were absolutely no breezes. I would drink water and rest for five minutes to give myself a break. Sometimes, I found myself looking up at the sky during those breaks, and I would declare to myself, that one day I would leave the farm and go somewhere I did not need to plow. I was busy all summer cultivating the crops and tending to the animals. My mother and sisters were also busy, while I was doing "the man's" work. They would use a hoe to chop grass from various crops, as I cultivated by plowing. They would gather blackberries, blueberries, huckleberries, and plums to make jellies and jams, and gathered vegetables for canning to provide food for us during the winter months.

One of the joys of my early years was Mama's homemade ice cream. Understand, that we had an old ice box. Twice a week during summer the ice truck would come by our house. Dad would purchase a 25 or 50-pound block of ice so that we had enough for ice cream and Kool Aid™ drinks. Mama would make vanilla ice cream custard on the old wood stove and add either strawberries or cherries. Then, my sisters and I took turns cranking the ice cream churn until the ice cream was firm enough to eat. On a hot summer day, with no air conditioning or fans, having something cold and sweet was absolutely a delight! We would sit on the front porch, eat our ice cream, and talk about events in our home and community. That is where most of my core values were formed.

At some point later we had access to electricity in our community. You can imagine that was a huge MOMENT for us! Having electricity meant we could also have an electric radio, rather than our old battery-operated radio and a refrigerator. One of our neighbors, Mrs. Lillian Conway (deceased), as I mentioned, had been one of our early elementary school teachers. She was fortunate enough to get a television when electricity first came to our community. Sometimes she would invite us over to see Friday night boxing matches with Howard Cosell as the sportscaster. We looked forward to that because that experience of seeing other people in a different kind of world inspired us to also want a different kind of life. Later we too were able to afford a black and white television that had three channels, ABC, NBC, and CBS. By the time we got a television, I was in high school.

A few of my classmates would venture to northern cities, such as Chicago, St. Louis, and Detroit during the summer months. When they came back home, they talked about what they experienced during their vacations. The experiences they shared with me motivated me to want a different life. I did a lot of introspection and became determined to search for ways to have that life.

Christianity was central in my home. My parents established a spiritual foundation for us. There were several ways they did this. First, they lived godly lives as examples for us. They loved us and taught us to love each other and our neighbors.

Second, they were leaders in our local congregation. Dad was our Sunday School Superintendent for forty years and served on the trustee board. He also took the leadership in the maintenance of the church. Music is a natural trait in our family. In fact, my oldest sister, Myra, learned to play an old organ that we had inherited from our grandparents. She played by ear. Mama had a great singing voice and would sing hymns and spirituals in the fields, while cooking, doing dishes and cleaning the house. So, naturally she would sing in our church choir, and encouraged us to sing in the choir.

I loved going to church. I guess it was because I had my first cousins about my age that I played with on the church grounds. Our church had summer revivals which lasted five nights, Monday through Friday. The front pew, on the right side of the church, was designated as the "mourner's bench" during revivals. The mourner's bench was the place where unsaved persons or persons who had not accepted Jesus Christ as their personal Savior sat during revivals. The main purpose of the evangelist was to get those people converted so they could make a public confession of Christ and be baptized. Although we were Methodist and had been baptized as infants, when we were of age, roughly 12 years old, and made a public confession of our faith, we chose to be baptized again.

If you were "unsaved" or a sinner, the evangelist and members prayed for you every night believing you would finally profess your faith. None of us "unsaved" people wanted to be sitting on the mourner's bench on Friday night. The primary reason was that the service would not end until every one of us would confess our faith. I had watched the dynamics of other revivals, and I knew, come hell or high water, I was not going to be on that bench on Friday evening! I knew that if anyone was on the bench on Friday, the evangelist was going to do everything in his power to get him or her saved. He was going to start with hellfire and brimstone and SCARE the hell out of us, to get everyone into the kingdom. So I decided I was going to give my confession on Thursday night. After the sermon on Thursday night I stood up, during the invitation for discipleship, walked up and shook the evangelist's hand. I think I was the last one on the mourners bench that day. Even though I attended church and participated as my parents requested, I did not experience a saving faith until years later.

As I mentioned earlier, our old home was located about two miles, back in the woods, on a dirt road. Mama always wanted to be closer to the highway. After daddy got his degree, his income increased. With the added income, sales of cotton, cattle and timber, and the assistance of family members who were carpenters, dad could afford to build us a brand new three-bedroom, ranch style home. The house was located about a half mile from the highway. We were all happy, but especially mama! She enjoyed being a homemaker in the new home and did not work outside the home until after I finished high school.

One day after Sunday School, an older first cousin, Robert Beamon, Jr. (deceased) and I were playing on a pile of gravel in front of the church. One of our uncles approached us and I remember him saying, "One of you boys is going to preach one day." We each looked at each other, pointed a finger at each other, and said it's going to be you." Little did I know how prophetic that statement would become later in my life.

Chapter 2

EDUCATION IS ESSENTIAL

"An intelligent heart requires knowledge,
and the ear of the wise seeks knowledge."

PROVERBS 18:15 (ESV)

Tougaloo Southern Christian College, as it was called during those days, was a small private Black school located near Jackson, Mississippi. It is currently a Historically Black College (HBCU). Tougaloo had its origin in 1869. It is affiliated with the Disciples of Christ and the United Church of Christ. Some of its professors were from Ivy League Universities. It has always been a prestigious school, primarily for Black students. I graduated in 1957 from Camden High School. Mr. Avaarah Strickland (deceased), my high school principal, convinced my dad that it would be a good idea for me to attend Tougaloo. He must have seen some potential in me that he felt could be developed in a school such as Tougaloo. He took a special interest in me by giving me advice and making suggestions about school during my senior year. He was an alumnus of Tougaloo and apparently felt that my potentials and the school would be a good match.

My Mother and Father at their 50th Wedding Anniversary

Both of my parents were strong advocates for education, especially with dad as an educator. We were told that education was the key to a potentially good life for ourselves and our future families. As with most teenagers, this did not strike me deeply at first. But when I went off to college, Mr. Strickland reinforced my parents' philosophy about education. I could see that he really wanted me to succeed. To encourage me to strive harder for good grades, he set up a running account at Moman's Grill, located just outside of the gate of the college.

I could go there as I desired to purchase burgers, soft drinks, and other snacks. He made no demands, but I knew that this was his way of encouraging me to do well in school.

When I got started at Tougaloo I didn't have a clue as to what I wanted to do in life. We did not have guidance counselors or Aptitude tests in my high school.

I decided to start out in an area I thought would be easy, so I registered as a history major. My professor was Dr. Ernest Hall (deceased), a White paraplegic who knew his subject well, but his monotone lectures bored me stupid. I had to fight through each class session to stay awake. I could not wait to change my major from history to biology during my second semester.

My change to biology was a breath of fresh air. My professor was Dr. Charles Arzeni (deceased), a brilliant White professor from South Africa. I looked forward to attending his class sessions because he was energetic, humorous, and very likable. I decided this is where I needed to be. To make the classes more fun, my roommate, Don Jennings (deceased) also majored in biology.

Learning experiences do not always come from the classroom, they also come from real life experiences. The one that I am about to share happened during the first semester of my freshman year. To think of it now is humorous, but at the time it was the most embarrassing experience in my young life. The 4-H Club (Head, Hands, Heart, and Health) in our county held an annual banquet for its leaders and members.

One of the main goals of 4-H Club work was to learn how to farm more productively, including better yields of cotton and corn, as well as better cattle production. It also included food preservation and other ares of farm work. The logo was a green four-leaf clover. The club represented practical and hands-on learning, which came from the desire to make public school education more connected to rural life. During my high school years I was active in 4-H Club work.

The Madison County Agricultural Agent, Mr. Thomas Mackey (deceased) asked me to be the keynote speaker at the event. A keynote speaker is asked to deliver a message that is inspiring, informative, and motivating. The speech is typically the opening or closing of an event and sets the tone for the rest of the event. I had never spoken to a group of any size before. However, because I was in college, and was asked to do such a significant thing, I thought it would be a piece of cake. I had gone over in my mind what I would talk about and decided that I could wing it without putting my ideas on paper. I felt that I could talk about some of my experiences while back in high school on the farm.

On the night of the event I rode with another student to Canton, who dropped me off where the event was held. I had never been to the place before. I did not know anything about the audience, or who would be attending, other than some teenagers from high school. Upon my arrival, dad and Mr. Strickland were already there. When we walked in, I was not prepared to see that three fourths of the people there were White parents and their teenage sons and daughters and city officials, the rest were Black parents and their teenagers.

I was very nervous and knew that I was unprepared. I knew that I was to speak about 5-10 minutes. I had been involved with methods of increasing the yields of cotton and corn by using learned principles from 4-H Club's tested and tried results.

Mr. Mackey gave me a great introduction. I came to the podium, with all eyes fixed on me and began my speech with these words; "Thank you Mr. Mackey for your kind words. 4-H Club work is efficient." Then I repeated those same words, but at a higher pitch in my voice, "4-H club work is efficient!" All my thoughts just simply vanished. There was absolutely nothing else in terms of the subject matter that came to mind!

I could just feel my dad and principal praying and pulling for me, but to no avail. I then concluded my short 1 minute (whatever) by thanking the bankers and all those who had made the banquet possible. But then, having embarrassed myself, my dad, my principal, and Mr. Mackey, I had to return to my seat next to dad. Both he and Mr. Strickland had their heads down as I came back to my seat. It was a painful experience. It was a tough act for Mr. Mackey to follow, especially after he gave such a glowing introduction. As best as I recall he said something to the audience like, "Well, that was short and sweet and now you get to go home early!" It was an experience that I will never forget, one which haunted me for years!

After everything was over that evening, Mr. Strickland pulled me aside and gently said "Walter, from now on, whenever you are asked to speak, get some note cards, and make an outline. Write down some key words and thoughts, even if you think that you don't need them. You will feel a lot more confident and comfortable just knowing that you have them if needed." That sage advice from more than sixty years ago has stuck with me ever since.

That night I had to ride with dad, 13 miles back home. Thankfully, I do not recall any conversation between us during the drive. He was more embarrassed than I because it was a stain on the Beamon name. Dad was well known in our community and well respected. Dad could walk into the banks in Canton and make a loan with only his signature. He did not have to use his property to secure the loan. Dad had the credibility to make the loan with a handshake, because they knew him to be a man of his word! This was during the 40s and 50s.

I now look back on that night of horror, that painful experience, as a blessing in disguise. As a schoolteacher and certainly, as a pastor and chaplain, I have had to do a tremendous amount of public speaking. That mistake, in hindsight, formed my development for public speaking. I never forgot Mr. Strickland's, (later Dr. Strickland's) words of advice.

God really does work in mysterious ways. Since that time as a 17-year-old boy, I have always tried to make necessary preparation when speaking publicly.

The second learning life experience that was pivotal in impacting my adulthood, happened during my sophomore year, which was less than a year after my implosion at the 4-H Club banquet.

At the beginning of the semester, all students had to have physical examinations administered at the college. The school physician took basic vital signs by listening to your heart and lungs, and taking your blood pressure and temperature. I thought I was in great health; but when he completed his examination of my heart, he looked at me with a serious expression. I recognized that he was hesitant to speak. My mind started twirling with all kinds of thoughts. I could not imagine what he was about to say to me. He then told me I had a very erratic and irregular heartbeat and needed further testing. He referred me to a cardiologist (heart specialist) in Jackson. He informed me that he would have them to set up an appointment as soon as possible, as this was a significant problem.

I was an 18-year-old kid. Hearing this information greatly impacted my psyche. I was devastated and unsure what my future held. To me, having a heart condition was tantamount to having a death sentence. It affected me so much that I was unable to sleep at night and I curtailed my physical activities. The problem for me was that shocking news, which seemed to exacerbate at night when I was quiet. My heart would palpitate, and I was literally scared out of my wits! I was afraid to go to sleep, for fear that I would not awaken.

Seeking solace and medical intervention, I consulted with a specialist who prescribed Digitalis, a medication that would become my constant companion for the next four years. Amidst this turbulent period, a glimmer of relief emerged. The specialist reassured me that I need not fret about the looming possibility of a military draft, a prospect I dreaded. I did not want to be drafted.

All males turning 18 were required to register for military service with the Selective Service System. I had to go through an extensive physical examination. Unlike my contemporaries who expected to pass the physical, I was not worried because of my "heart condition." Well, unbeknownst to me, the examination showed that I did not have a serious heart problem as previously told. I had an irregular heartbeat, but it was not serious. I did, however, continue taking Digitalis for several more years because it seemed to stabilize my heartbeat. I met all the requirements for the draft. However, my role as a science teacher held a silver lining: educators in math and science were eligible for deferment, sparing me from an unwanted call to service. The highlight of 1959, my sophomore year, was where I pledged to and became a member of the Gamma Upsilon Chapter of Alpha Phi Alpha Fraternity, Inc.

The early 1960's was a period of great turmoil and unrest because of the Civil Rights Movement in the deep South. The lunch counter sit-ins in Jackson and the activities of the Freedom Riders from the North brought great fear to people of color. Many of my classmates participated in the lunch counter sit-ins, some of these took place only a few miles from our campus. These students were bold and courageous. Most of them were abused and arrested. Although I was not among them, I fully supported their activities. These young men and women put their lives on the line to focus on activities designed to challenge and change the "system" for equality and justice for Black people in this country.

The climate in the deep South prevented many Blacks from registering to vote. My dad desired to register and vote, but felt his teaching job would be jeopardized if he did. In 1963, my sister Myra was teaching and supporting the causes for justice for Blacks. She felt it was imperative that all Black people should have the opportunity to vote.

She convinced Dad to go with her to the Madison County Courthouse in Canton to vote. She and Dad passed a bias test, and after each of them paid a $2.00 poll tax, they were allowed to vote.

Chapter 3

THE LOVE
OF MY LIFE

Then the Lord God said, "It is not good for the man to be alone.
I will make a helper who is just right for him."

GENESIS 2:18 (NLT)

In the fall of 1960, my senior year of college, I met the love of my life, Ikie Jean Haynes! In October, the school held a Halloween dance in the gymnasium. The Disc Jockey (DJ) played records that students danced to and invited some students on the stage to sing. There was one student in particular who could sing like Brook Benton, a well-known Black pop artist. He had a crush on Ikie. I had seen Ikie on campus before, and I liked her too. Both of us danced with her. However, with fast paced music my rhythm was limited, in other words, I had two left feet. He could dance fast and slow. Consequently, he danced with her more. However, when slower paced music was played, I beat him to her. She did not appear to be very interested in either of us. At that time Ikie was a "book worm," and very studious. Toward the end of the evening, I made sure I was near the door when she started back to the dorm. I asked her if I could walk her home. She agreed and that was the beginning of our courtship..

On Valentine's Day, I did not have money to buy her a proper gift, so I made an 18-inch alligator out of latex rubber. I used an alligator mold from the various molds in our biology lab. I painted it green and tied a pink ribbon around its neck. I painted the initials for my name, "WEB" on its side. I was embarrassed to give her an alligator as a gift. What would she think of me? My roommate thought it was hilarious, but I gave it to her anyway. Ikie loved it or did a darn good job of pretending. WEB's colors were her sorority colors. I knew then that she would be the right partner for me as I faced future challenges. She kept WEB for many years after our marriage.

Ikie came to Tougaloo with the express purpose of becoming a medical professional, either a physician or a registered nurse. Tougaloo had a pre-med program for students who were planning to become medical doctors, and an affiliated nursing program for students who planned to become registered nurses. The basic course requirements were very similar. Ikie has always been very disciplined. She had mentioned that at one time she toyed with the idea of becoming a doctor. However, when she realized how long we would have to be separated for her to become a doctor, she decided to pursue nursing instead.

In those days, in Mississippi, there were no schools for women of color interested in becoming registered nurses. Therefore, she had to leave the state to complete her education. Homer G. Phillips School of Nursing in St. Louis, Missouri, and Tougaloo College had established an affiliated program which provided a way for women to receive a Bachelor of Science in Biology and become a registered nurse with the five-year affiliated program.

When Ikie went to St. Louis to continue her education, I started teaching high school biology at Rogers High School in Canton. We were separated geographically by 700 miles for three years. I often wondered if our relationship could last, but I relied on an old saying I had heard: "Absence makes the heart grow fonder." There were few phones during those days, and they were pay phones in telephone booths. We communicated by writing each other. I looked forward to her weekly letters. I admit that she wrote me more frequently than I wrote her. That does not mean that I did not think of her often, I simply was not disciplined enough to write as often as she did. Occasionally I saved enough coins to call her on the pay phone, and it was a thrill to hear each other's voice.

Ikie upon graduation as a Registered Nurse

At some point during our communications, we talked about marriage. She made a statement I shall never forget, and my response would come back to haunt me later. She said, "I don't want to be married to a doctor or a minister." I asked her why those two professions? She said, "A doctor or minister's time is not their own time. Their time is obligated to their patients or their parishioners. There is little time left for family." I remember emphatically responding, "Well, baby, you don't have to worry about me being either of those, because I am going to remain a biology teacher." I really thought at the time that I would always be a biology teacher. But no one foresees the mind of God! *"God's ways are not our ways and His thoughts are not our thoughts."* (Isaiah 55:8-9) After that discussion, I enthusiastically worked in my career field.

My starting teaching salary for the eight-month school year was $2,900.00. That equated to approximately $363.00 per month, which did not go very far in those days. It was a financial challenge, but I bought my first new car.

It was a 1962 white Chevy Impala, 2-door hard top, with red interior. I thought it was a "BAD" car. It was a head turner. About three times a year I put my new car on the road to visit my sweetheart. Most of the time I would take off on the weekend and leave early enough on Sunday to be back in the classroom on Monday morning. At the age of 22 you can do anything.

We continued long distance dating for two years. At the beginning of her third and final year in St. Louis, we were married on August 10, 1963, on her parent's lawn.

It was a beautiful wedding, but a hot day with the temperature at 95 degrees and about 98% humidity. I did not know if I was about to pass out because of the weather or because I was making a lifetime commitment! Getting married on a hot day, on a lawn in August, in Mississippi, is no joke!

Our Wedding

I was all set to bring Ikie back to Mississippi to start our lives together; but when I mentioned my plans to her, she promptly refused. "Absolutely not, she told me. I AM NOT GOING BACK TO MISSISSIPPI now or ever!" I didn't understand why she was so adamant about not coming back. This was her home, where she grew up, and her parents lived here. I did not understand until she enlightened me.

She reminded me of the racial unrest in Mississippi, as well as other Southern states. People were being beaten, jailed, separated from families, and persecuted for simply trying to live a meaningful life with opportunities to be fully human, with equal rights. This was the case throughout the country, but Ikie had faced several racial confrontations in Mississippi that resulted in a deep anger and resentment that made her want to seek a future somewhere other than the state of Mississippi.

Ikie's mother, Lillie Mae Hall Haynes, of Jefferson Davis County, was also a teacher. She received her teaching certificate and bachelor's degree pretty much like my father did. To receive her monthly salary she had to make sure the absences and presences of each pupil she taught equaled the total number of school days for the month for each child. This tedious task was completed at the end of each month. Since many pupils had to miss during the harvest of various crops, it was very challenging for teachers to be accurate and often took several hours. For Ikie's mother this task was completed at the courthouse in Prentiss. Most of the time her mother left her and her sisters with an aunt.

One day her aunt was unavailable to keep Ikie and her two sisters, so her mother took them with her. The girls had never been in the courthouse before, although they had seen it from the outside and marveled at how "big and beautiful" it was. They also knew that there was a toilet outside the courthouse in the back for Black people to use because they had occasionally used it. When they entered the courthouse that day, they observed the beautiful shiny floors and benches and tables.

As Ikie (about nine years old) skipped down a hallway with her mother and sisters she also saw a sign that said, "Women's Restroom." Her mother got to work getting her roll tallied. On one of those long beautiful shiny tables. Ikie and her sisters sat at the table and played dot-to-dot to pass the time as their mother worked. After a while Ikie's second oldest sister (Barbara) needed to use the toilet and said so aloud. Hearing her, Ikie said, "I know where it is." She had made a mental note of its location while entering the courthouse. Her mother, absorbed in her work and not thinking they would choose to use "the pretty women's restroom," did not go with them.

When they got to the "pretty women's restroom" they saw a white man sitting nearby. As mentioned before, they had not been in the courthouse before and because they had been shielded from a lot of racial encounters, they did not know they were not supposed to use that restroom. They had just finished using the restroom and commenting on how soft the toilet paper was and how "pretty" everything was, when the door busted open and seven white men pounded into the room carrying guns. One of the men shouted, "WHAT IN HELL YOU GODDAM NIGGERS DOING IN HERE?" Other men were also shouting. Ikie could not remember what they were saying, but there was a lot of commotion and she and her sister were frightened out of their wits. The old man who had been sitting nearby when they entered was with them. Their mother heard the commotion and came running to their rescue.

Understand that, Mrs. Haynes, my mother-in-law was a proud and proper speaking woman. She taught her girls, "ladylike" behaviors and speech. This was so much a part of who she was that even after we were married and living in other states, Mother Lillie, as I fondly called her, would correct, by mail, some grammatical errors that Ikie had made on a previous letters she had written her mother. On this day, Mrs. Haynes became a different person. Ikie heard her mother say to these men, "Mista, dey didn't mean nuthin. Dey just chillun and did'nt no no better." One of them asked where Ikie and her sisters were from. Mother Lillie said, "dey from hea but they never been in dis place befoe and dey didn't mean no harm!" Fortunately, she had completed her work, and they were allowed to leave.

That day changed Ikie's perception of her world. Up until that time, her parents had limited the environment of Ikie and her sisters primarily to home, church, and school. Ikie saw her proud intelligent mother become someone she had never seen before. She became subservient to a group of ignorant people in order to save their lives!

These men were imprisoned by their long-time belief that they were superior to these NIGGERS who did not belong in that place and they were "in charge." To Ikie, that day began a long-term hatred of White people. At age nine, she did not understand.

When Ikie traveled back and forth from Prentiss to Jackson while studying at Tougaloo, she had to travel by bus. There were not many buses that made the 2-hour run each way from Jackson to Prentiss, and back to Jackson. The bus would leave Prentiss practically empty, but stopped several times to pick up passengers in other small towns enroute to Jackson. Often the bus would become crowded by the time it reached Jackson. On several occasions, she was comfortably seated on the second or third seat when a couple would get on the bus and the only seats available were in the back of the bus. The bus driver would point at Ikie and say something like this, " Hey you, you got to go back there and take a seat. These people need that seat." Ikie's parents had told her to always move when that happened, and she always did. Those experiences were never forgotten, instead they were seared into her psyche. God works in mysterious ways and helped her work through her conflicts through a friendly White nurse whom she encountered at Tougaloo. However, these intense, severe experiences prohibited her desire to return to the state. After that conversation about coming back to Mississippi to live and learning why she had refused to come back, we decided to move to Chicago, Illinois. That was the easiest decision because we both had friends living there.

Chapter 4

CHICAGO — OUR NEW WORLD

"Do not be afraid or discouraged, for the
Lord will personally go ahead of you. He will be with you;
He will neither fail you nor abandon you"

DEUTERONOMY 31:8 (NLT)

In 1964 I applied to the Chicago Board of Education for a teaching position and was hired to teach high school biology at Tilden Technical High School, on the South side of Chicago. In contrast to Canton, where my salary started at $2900.00 per school year, in Chicago my salary started at almost $6000.00 a year.

When we went to Chicago, everything we owned could fit in our 1962 Impala. We had no money, and we could not afford our own apartment. The friends we knew in Chicago were friends we had met at Tougaloo. They were gracious enough to allow us to stay with them for a couple of weeks until I received my first paycheck. While we were staying with them, we found a small, one-bedroom apartment that we could afford. We borrowed a single roll-a-way bed and a card table with four chairs from our friends. These simple items, along with the stuff we brought in the car furnished our small apartment located at 8812 S. Cottage Grove. Ave. on the Southside of Chicago for several months.

I had a rude awakening the first day I arrived at my new place of employment. When I drove up to the campus and parked my car, I saw several police cars there. I wondered what policeman were doing at the school. As I approached the entrance, I saw the policemen searching students for weapons, paraphernalia, and drugs. This was shocking to me because I had never seen anything like this back in Canton. I admit the revelation of policemen searching students was unsettling. I found myself feeling anxious and a bit uneasy.

Tilden Tech. was located on 47th & South Union Streets. I later learned that this was one of several locations for street gangs. The Crips and Bloods were rival street gangs that fought to control this area, as well as other areas in the nearby vicinity. Some of my students were either members of one of the two gangs or knew someone who was. Therefore, daily fights were expected. This little country boy, from the sticks of Mississippi, had no idea of what he would face in his new job when he migrated to the mid-west, the land of Lincoln.

Chicago was a busy, intimidating, large city. Learning to drive in the city was challenging, but I had to learn and learn fast. Ikie's job was across town on the West Side and we lived on the South Side. Ikie secured a position at Michael Reese Hospital and Medical Center, located in the Bronzeville neighborhood. With only one car, I often had to drop her off at work and /or pick her up. We soon settled into a routine of working all week and partying in clubs and bars on the weekends. We met new friends who also enjoyed "clubbing" or playing cards at someone's residence.

Despite the dangers associated with my job, I enjoyed planning my lessons and working with other teachers at the school. It gave me great pleasure to see one student learn something in my classroom, especially because of the environment in which they lived. We would take short weekend trips when we could. We visited our parents on holidays when Ikie could get away from her demanding job.

I found myself occasionally getting restless with our established routine. I decided to pursue fishing as a hobby. I bought good fishing equipment and would often spend entire Saturdays fishing. If one is not a true fisherman, fishing can be boring, especially if you do not catch fish. After several months, my enthusiasm for my fishing hobby began to lessen. Recognizing this, I decided that photography would be an interesting hobby. So, I pursued that for a while. I took pictures of beautiful nature sceneries and people. I took a course in photography, and learned how to develop film in our small bathroom. After a short period, I lost most of my interest in photography as well.

The move to Chicago is where my whole life would be turned UPSIDE DOWN or theologically speaking, RIGHT SIDE UP! I believe God provided a "divine appointment" with a man of God. A fellow teacher at my school was on fire for God. His name was William (Bill) Andrews. Bill taught chemistry. Since we were both in our mid-twenties and both taught sciences, we spent a good bit of time together. Over lunch or during lesson preparation periods, we would discuss various issues in our lives. I shared with Bill that I had found myself growing more and more restless. I had tried several hobbies, but still found myself unfulfilled. I still experienced a void in my life.

One day over lunch, Bill shared his faith with me. He had a powerful testimony of how God had changed his life. He was bold and courageous with his witness. When Bill spoke of his relationship with Jesus Christ,

I could sense his sincerity and commitment to his Christian faith. I listened to his powerful witness, and I longed to have a witness such as his. As we spent more time together, I noticed a gradual change occurring in me. At some point, I prayed inwardly for a strong relationship with Jesus Christ. I longed to have peace in my heart and a testimony to share. Bill introduced me to Moody Bible Institute, a radio ministry that was broadcast throughout the country.

I took advantage of the information Bill gave me about Moody Bible Institute and listened to scripture, music, and messages that filled the void in my heart. I found myself reading more materials about faith and worship. I discovered that there is a big difference between going to church as I did back home and really having a relationship with Christ. I prayed for a strong relationship with Jesus Christ, and it happened!

One day after lunch, I went back to my classroom and was suddenly overcome with a powerful flood of joy in my heart. I literally felt the Holy Spirit at work in me. I prayed to God to accept me as His child. I yielded my life to Jesus Christ. I was unable to teach my classes for the rest of the afternoon. I gave my classes assignments and removed myself to the adjoining biology lab. I sat with tears streaming down my face, for a transformation was taking place in my life. When I yielded myself to Christ, I felt something that I had never experienced before. I knew that it was God!

When school ended that day, I rushed to Sears, Roebuck to purchase a bible. I had a hunger and thirst for the Word of God. Each day after work I would come home, sit in my recliner, and read scripture for hours and hours. Sometimes I would read so long that I fell asleep and would wake up later around midnight. I knew that changes were occurring in my life. I found myself gradually losing interest in teaching. Prior to this time, I loved making lesson plans and observing learning taking place with my students. I loved seeing my students learning and excelling. Now I was losing that interest as my interest grew in spiritual matters.

Ikie noticed the changes taking place, but she was not too concerned because she had seen me change my interest several times before. So, she thought this change was just another fad and soon would fade away. As time went on, she noticed other changes. Whereas, previously on weekends we would go to night clubs, stay out until the wee hours of the morning, and sleep in on Sunday mornings, now things were different.

Back home I had been a member of the Liberty Chapel African Methodist Episcopal Zion Church (AMEZ). I had noticed a small church a few miles away called St Mark AME Zion Church and I made a mental note about its location. One Sunday morning I decided to visit St. Mark. I went alone, Ikie slept in. I felt comfortable and at home in the worship service and made a commitment that I would continue attending. During the service I recognized the organist from Canton whom I had met when I was teaching at Rogers High. What a small world! She and I connected, and she encouraged me to join their choir. Since I had sung in our choir back home, I felt joining the choir would help me to grow spiritually. I was excited.

When I got home, I told Ikie how much the worship service meant to me. I described the church and the people. Then I told her about the organist whom I knew back in Canton, and she was single. I told her that I had decided to join the choir. At first Ikie appeared to be disinterested. After I had been practicing and singing in the choir for several weeks, she told me, "I think I'll go with you, to protect my INTEREST, meaning she did not like the idea that I had known the organist before, and she was single. By going to church, bible study, choir rehearsal and Sunday school, we did not have time or the desire for night clubbing. So, we did not see our friends. One day Ikie asked me, "What are we going to do for friends?" I responded, "God will provide the right friends!" He did. After choir rehearsals on Friday evenings about seven of us would go bowling. We would also share meals in each other's homes. We got to know more about their backgrounds and their Christian journey. After 57 years, some of the friends God provided are still our friends as I write this book. We still communicate with each other.

Chapter 5

MY DRAMATIC
CALL TO THE
MINISTRY

"You did not choose me, but I chose you and appointed you that you
should go and bear fruit and that your fruit should abide, so that
whatever you ask the Father in my name, he may give it to you."
JOHN 15:16 (ESV)

After my conversion experience in the spring of 1965, I knew that God
was at work in my life. However, I was not sure what it was. I did have
the impression that my time in the classroom was limited. It was so
prevailing that one day I found myself praying for discernment. I
prayed for direction by saying, "Lord I don't know what you are doing,
but I know that you are at work in my life. I just need to know what
direction I am to go."

A few weeks passed and I received a call from my father-in-law, Mr.
Isaac J. Haynes (deceased). I had never received a call from him before,
period! This was so unusual because Mother Lillie usually called Ikie

and all of us would talk. So, his call to me was very significant. We talked for a while about the family, his crops and other small talk. I wondered what his real purpose was for calling. Eventually he paused and said to me, "Walter, you have been on my mind lately. I may not live to see it, but you are going to preach one day." It took a few moments for me to process what I had just heard. I did not know how to respond although I must confess that the thought of preaching had crossed my mind.

Ikie and me in our apartment.

About a week later, my brother-in-law, The Reverend Horace Buckley, who was the Senior Pastor of Cade Chapel Missionary Baptist

Church, in Jackson, called me. He said, "Your sister Myra had a disturbing dream last night. She saw your dad standing on a wagon, just preaching, preaching away. She woke up greatly perplexed, trying to figure out what the dream meant and said to me, "I don't understand that dream. Daddy is 65 years old, how could he be preaching like that?" I said to her that was not your dad, that was your brother!" Horace and I had never talked about me preaching in any of our previous conversations.

Having received both calls, I knew that God was leading me into the ministry. I did not tell Ikie or anyone else what was in my heart and mind. I prayed another prayer, "Lord with these calls, you have confirmed that you are calling me to the ministry. I am willing to go where you lead me. I believe I need a theological education in order to be effective in my ministry. Lord, I need you to open some doors to allow this to happen." I needed to find a way to tell her what I was feeling about "my call."

My sister, Agnes lived in Gary, Indiana, and was getting married in early June 1966. My mother came from Mississippi by train for the wedding. We met her at the train station in Chicago. Ikie sat on the back seat and Mama on the front passenger side. On our way home I softly mentioned to Mama that I felt that I had been called to the ministry Ikie, hearing my comment to mama, sat up in the back seat and asked, "What did you just say? You've been what? Did you say you've been called to the ministry?" I had to respond truthfully, "Yes I have!"

Within six weeks after this conversation, I received a letter from the Dean of the School of Religion at Virginia Union University located in Richmond, Virginia (later renamed Samuel DeWitt Proctor School of Theology). Getting this letter was shocking to me because I had not applied for admission. The letter was an invitation to attend the seminary.

We ignored the letter, because we felt that it would be impossible for me to attend on such a short notice. We had financial obligations.

I wondered how the Dean knew that I had been called to the ministry. I had been very guarded about discussing my call with anyone. I shared my thoughts about this with Ikie. She informed me that for several months she and a friend, whom we both knew, had been exchanging letters. I was unaware that they had been communicating. The friend was married to a former fellow teacher from Rogers High. Ikie had told her about the changes she had seen in me. Apparently, she shared this information with her husband, Emmett Burns (deceased), and he had informed the dean that I might be a candidate for seminary training.

Within two weeks I received a second letter from the dean. Ikie then suggested that out of courtesy I should at least respond to the invitation. She suggested that I could respond by mentioning some requirements which we thought might be more than the school would be willing to grant, especially for the current year of 1966. Our rationale was we would be buying time. Our requirements included: a full tuition scholarship, a household move at the school's expense and Ikie would need employment.

I received the third letter from the dean two weeks later. "Dear Mr. Beamon, I received your letter, and this is what we can do for you. We will provide a full tuition scholarship, however, you must maintain a "B" average. We will move you at our expense and I am sure that your wife would be able to find work nearby."

This floored both of us. WOW! This is not what we expected. We had hoped that the dean would have focused on the next year, 1967. I completed my application which included questions about my spouse. In astonishment, Ikie became defensive and blurted out, "How do you know you have been called anyway?" I knew she was thinking about all the ramifications this move would entail.

Ikie was thinking about all our bills and the fact that there would be only one income in our household. Her response perturbed me because she was questioning my call. I responded, "Why don't you pray and ask God yourself and see what He says." She was too stubborn to pray. Ikie didn't really want to leave Chicago because she had dreamed that one day we would purchase property there and make "big bucks." But the thought crossed her mind that if she could get a job in the medical infirmary as an additional nurse, perhaps that would work. She dismissed that idea because she surmised that most colleges and universities only had one full time nurse.

Ikie was home alone a couple weeks later when the doorbell rang. The postman delivered an air mail special delivery letter from the President of Virginia Union University addressed to her. The letter offered her a full-time position as the second nurse in the infirmary. She was shocked beyond belief because she had not applied for nor prayed for the position. Two crucial points must be made. First, Ikie had never applied for the position as a nurse. Second, this was the first time that the University had recognized the need for a second full-time nursing position! So, the question must be raised, was this purely coincidental or what? This was the turning point of her acceptance of my call and she no longer had doubts of its authenticity.

After the shock of the letter, Ikie knew that this whole dramatic period had to be of God, so she slowly began to focus on our future ministry and all that would later unfold. She never again asked me the question, "How do you know that you have been called to preach anyway?"

At this point we only had three months to get prepared for the move because school would start again in September. We had to arrange and schedule the move with a moving company along with other things associated with the move. We had just purchased a brand new 1965 Buick Electra 225, prior to knowing we would be moving. One day Ikie was driving along the streets of Chicago, when suddenly, a man driving a truck behind her plowed into the back of it damaging the rear end and trunk. He pleaded with Ikie not to call the police because he would lose his license and other sob stories. He assured her that he would pay for the repairs and gave her his name and address. I was driving the city bus for summer employment at the time. When I came home, she was literally in tears about the accident. I went to the local police station to report the situation. They indicated that there wasn't much they could do inasmuch as they were not called at the time of the accident.

They did, however, take down the information on his name and address and checked it out only to discover that he had given the wrong address, but he was actually living next door to the wrong address he had given Ikie. In the meantime, I went to a General Motor's dealership and explained my dilemma of needing to get the car fixed within the next few days. He was very sympathetic with me and said if he could find the parts needed, although he was doubtful, he would put my job ahead of others to get me going. He called me the next day saying he was very surprised to find everything needed in a warehouse and he would be able to get started immediately.

But before I had received the call from the dealership, I questioned God as to why He would allow that to happen to us, knowing that we were going to Virginia to do His will? Long story, short, the man who had caused the accident came to our apartment with a brown paper bag with $650.00 in cash, the cost of the repairs of our car. When he came into our place, I had a sense of compassion for him and my mind changed to focusing on his salvation. I think I prayed for him on the spot. The other miracle was a test of our faith. Our car was repaired and ready for pick-up two days before we were scheduled to leave Chicago.

God is worthy to be praised! He knew that our intentions were upright and our motives pure, so He blessed us with a testimony of His goodness and grace. There are many other incidences that happened that we knew God was at work in the matters. Thanks be to God who honors all His promises.

We left Chicago enroute to a city we had never seen before, not knowing what to expect in Richmond, Virginia. We had many questions on our minds, questions such as, how will we be accepted there? What does the Lord have in store for us? These questions made us a bit anxious as we traveled to Richmond.

Upon our arrival we felt some comfort because we had communicated with our friends there and they had invited us to stay with them until we were able to find an apartment. They were very gracious hosts. It was urgent for us to find an apartment quickly because our household goods would be arriving within two to three days after our arrival. We started looking for a place the day after our arrival. It was important that we were located very close to the campus because of Ikie's job. We found apartments within walking distance to the campus, so we stopped and inquired as to availability at the housing office but were told that none were available.

I didn't tell them that Ikie had a job at the University. I mentioned that to our friends, and they indicated that I should go back and let them know that I was a student, but my wife would be employed at the University. When I did make it known that Ikie would be employed, suddenly apartments were now available.

They had apartments available all along, however they were careful because they had trouble collecting rents from students. We were able to rent a very nice apartment that same day which was in close proximity to the campus. The movers came to deliver our household goods the very next day. We had been fearful that they might have to be put into storage. What was further amazing was the fact the wall-to-wall carpet custom made draperies we had brought from Chicago fitted the new apartment as though they were made for it! Everything just fell into place. That alone was further confirmation for us that it was a God-Thing! God is good all the time! All the time God is good!

The words from John's Gospel, chapter 15:16 took on new meaning for us;

> *"You did not choose me, but I chose you and appointed you that you should go and bear fruit and that your fruit should abide, so that whatever you ask of the Father in my name, He may give it to you."*

Chapter 6

SEMINARY — A NEW VENTURE IN LEARNING

"All scripture is inspired by God and is useful to teach us what
is true and to make us realize what is wrong in our lives. It corrects
when we are wrong and teaches us to do what is right. God uses
it to prepare and equip his people to do every good work."

2 TIMOTHY 3:16-17 (NLT)

The first part of our journey of faith was now complete. Everything was new to us. It was a different environment and we felt akin to a fish out of water.

In this region of the country the accents of most people we encountered were very different. Although the seminary was a member of the Baptist General Convention of Virginia, the students came from a wide variety of backgrounds and denominations including Methodist, Pentecostal, and Baptist. I was excited to study in an environment so diverse in denominational backgrounds. I felt it was a blessing that God had provided this opportunity to be a more effective minister due to this learning experience.

While most of my classmates aspired to become effective pastors for their future congregations, there was one standout among us – Robert Jemerson from Dallas, Texas. Robert had a unique goal: becoming an Air Force chaplain. It was uncharted territory for someone like me, who had limited exposure to the military world. My only connection to the military was through an older cousin. With unwavering conviction, Robert tried to persuade the rest of us that chaplaincy was the path worth pursuing. However, at that point in time, becoming a chaplain wasn't on the radar for my classmates and me. Our singular focus was to prepare ourselves to pastor a church.

Once we established routines everything seemed to fall in place. Since it had only been five years since I had completed my undergraduate work, it did not take long to get back into the swing of an academic setting. I found it necessary to spend hours in the library. Thankfully, I did not have to work and could afford to spend the time needed. Our income at that time was cut less than half of our Chicago income.

Ikie was blessed to find a part time job in the mornings from (8 a.m.–12 noon) to supplement her full-time job from 4 p.m. to midnight.

In seminary, I met an older gentleman in one of my classes. He was approximately 65-years old and working on his theological degree. He was the Reverend J. Oliver Hart (deceased), pastor of Hood Temple AME Zion Church in Richmond, Virginia. This was a very prominent AME Zion Church. I was overjoyed to find a pastor of my AME Zion Denomination in classes with me.

Hood Temple proved to be a major center in our lives. Dr. Hart and his congregation were like family to us. Ikie was involved in the church and had a great relationship with his wife, Mrs. Lois

Hart(deceased). Mrs. Hart had been a pastor's wife for many years and shared much wisdom about being a pastor's wife to Ikie. The wisdom she provided proved to be vital to us throughout my ministry as pastor of church and as a chaplain in the Air Force.

During the time I spent with Dr. Hart, I learned that Hood Temple was founded during 1916. It was originally built as the Clay Street Methodist church. The original structure was erected in 1854 by Albert West, a notable Richmond architect. Dr. Hart also informed me that the AME Zion Connection was called, "The Freedom Church" because of its involvement in freeing black slaves. Frederick Douglas and Harriett Tubman, both abolitionists, were members of the AME Zion Connection.

Meeting Dr. Hart and developing a close relationship with him and his wife was confirmation to me that God had led us to Richmond. Dr. Hart became my "father" in the ministry. As a young man, starting out in ministry, he nurtured me and gave me opportunities to grow. He encouraged me to attend various denominational conferences and helped me prepare for my ordinations as a Deacon and as an Elder. Dr. Hart was the pastor of a large congregation while engaged in a full-time curriculum in seminary. I found him to be a saintly man, humble, and devout in his faith. I felt honored and blessed that he took the time from his very busy life to help me learn and grow to minister to God's people.

Deacons are ordained Methodist clergy who can lead by preaching, teaching, nurturing spiritual vitality, and leading ministries of service, love, and justice. The primary difference between a Deacon and an Elder is that a Deacon cannot bless communion, and an Elder can. Elders are concerned with the spiritual welfare of the congregation, while Deacons are primarily concerned with the physical welfare of the congregation.

My ordination as a Deacon in the AME Zion Church occurred during my first year of seminary in 1967. I was ordained as an Elder in 1968. Shortly after my ordination, the Presiding Elder/District Superintendent of the Petersburg District assigned me to a small church called Taylor's Chapel, located near Merdithville, Virginia, about 60 miles south of Richmond. Due to the extreme shortage of pastors, the Presiding Elder had been serving this church. It was a very small congregation with approximately 10 adults and 4-6 children. The membership consisted of farmers and two elementary teachers. The people would meet weekly for Sunday School and once monthly for worship service.

Business meetings were held on Saturday evenings on the weekend of worship service. I was excited to be the pastor of my first congregation. I was also apprehensive. This was a totally different role from teaching. The church members would look to me for leadership and spiritual guidance. Ikie and I would drive to the community on Saturday afternoons and would be invited to spend the night in the home of one of the families. After the first year I was assigned to a second church, St. Paul AME Zion, with about twenty-five members in the same general area.

The Fall of 1968 marked the beginning of my senior year of seminary at Virginia Union. I graduated in the spring of 1969 with a Master of Divinity Degree (M. Div.) I was the recipient of another scholarship that focused on a Master's degree in Pastoral Theology from Union Theological Seminary (Presbyterian). I graduated from that program in 1970 with a Master's Degree in Theology (M. Th).

Chapter 7

NEW SENIOR PASTOR— NEW CHALLENGES

"Go ye, therefore, and teach all nations, baptizing them in the name of the Father, and of the Son, and of the Holy Ghost: teaching them to observe all things whatsoever I have commanded you: and, lo, I am with you always, even unto the end of the world. Amen."

MATTHEW 28:19-20 KJV)

After my graduation, I switched my focus from academia to searching for employment. I really had no clue what God wanted me to do. At that time, I felt my realistic choices were to become a full-time pastor or to teach in a seminary. Given my background in teaching, Dr. Hart suggested applying to Hood Seminary at Livingstone College in Salisbury, North Carolina. That idea did not really appeal to me, however I followed his advice and applied for a teaching position. I was interviewed and offered a contract but declined signing it that day. I wanted more time to evaluate my options. In the meantime, Rev. Raymond Hart (Deceased), Dr. Hart's youngest son, called me. He knew that I had completed seminary and was seeking employment. He strongly encouraged me to call the Bishop over the Georgia Conference, Bishop William Foggie (Deceased), regarding a church in Atlanta, Georgia that needed a pastor.

My graduation from Seminary with my Dad

Ikie and I weighed the pros and cons of teaching versus full time pastoring. Ikie felt I should pursue teaching since I had experience in teaching. For some reason, I leaned more toward the church in Atlanta. I was very hesitant to call the Bishop.

There were several reasons for my hesitancy. First and foremost, we had never met before, and I had minimal pastoral experience. Raymond (Ray), who became my close friend, told me that the Bishop was a caring and "easy to talk to guy." I called the Bishop, introduced myself, and shared with him information about myself. I told him that I had recently graduated from both Virginia Union and Union Theological Seminaries. I had also been interviewed and offered a position on the faculty of Hood Theological Seminary in North Carolina. I still had the unsigned contract because I wanted more time to evaluate my options.

The Bishop indicated that he believed it would be a mistake going to Hood Seminary due to my limited experience as a pastor. He mentioned that he had a church, Shaw Temple, in Atlanta, Georgia, where he needed a pastor. He needed to make sure that he sent the right person.

And he said, "The more I talk to you, the more I feel you are that person." He offered me the church but admitted that the church had experienced some major challenges.

I requested the opportunity to visit and preach at the church before I made a decision. He approved and made arrangements with the officers of the church for our visit. He stated that he would mail the Certificate of Appointment that same week.

Approximately three weeks later, Ikie and I made the trip to Atlanta for the weekend. We were eager to see what the church was like. Both of us were apprehensive since the church had "major challenges." The church was located in Southwest Atlanta, about five miles from downtown. When we entered the property, we saw a large building with a huge sanctuary, and attached fellowship hall. There was also a ranch style house located about 100 yards from the church which had served as the parsonage.

We arrived in time for Sunday School, and met some of those gathered. An official showed us the sanctuary and Sunday School classrooms downstairs. We found the people very friendly and seemed eager to meet us. The officers showed me to the pastor's study, which allowed me time to prepare for the worship service.

Worship service began with about 25-35 adults, including a few children in a sanctuary that could seat 600 people comfortably. Mrs. Annie Sims (deceased), the organist played some very sedate music for the prelude to worship. I began the service by introducing myself and Ikie, from Richmond, Virginia, as the newly assigned pastor and first lady, by Bishop Foggie, the presiding prelate of the Georgia Conference. Then I proceeded with the order of worship. Everything went well with the service and one young adult responded to the call to discipleship at the end of the sermon. Later, I was told that no one had joined the church in months before then. That was a moment of encouragement.

Immediately following the worship service, I met with the Steward board to gather information about the financial standing of the church, the pastor's salary, and the need for the church to move us from Virginia to Atlanta. I was immediately made aware that the church coffers were almost empty with the church's bank account being frozen. This was not a good sign. This was the first evidence of the "major challenges" the Bishop had made reference to earlier. I inquired as to the salary of the pastor, which I already knew from my good friend Raymond Hart. I was informed that the salary was $115.00 per week, but they were quick to say, "But Reverend, we're not able to pay that now because we just don't have it. Our attendance and offerings are down!" My response to the board was, "My goal for the church is to have it grow and become self-sustaining. In order to do that, I need to devote my full time to the growth and development of the church.

Therefore, I will need a minimum of $115.00 per week and the parsonage renovated so we can live in it." With that being said, I encouraged them to continue to meet to determine how they would move forward in meeting our needs. They should let us know within a week. They called within three days to inform us that they had found ways to meet our needs.

My father in the Ministry, Dr. J.O. Hart invited me back to preach.

When we arrived back in Atlanta, the parsonage had been renovated, and I enthusiastically began my work as a senior pastor at Shaw Temple. Our first year was tough. I had to learn quite a bit as I hit the ground running. I had to get an understanding of some of the sticky issues that were still active as it related to my predecessor and his administration. I had to get to know the official officers of the church, the "movers and shakers."

I had to prayerfully consider many things as I navigated my way toward making appropriate changes. My first initiative was visiting the sick and shut-in members, and meeting their families. The second initiative was to visit members who were no longer attending worship services. Through these visits I gathered important information which helped me to make crucial decisions.

A statement made by Dr. Hart was etched in my brain and consciousness. He said, "Son, if you are going to be effective as a pastor in the Methodist Church, you must be a shepherd and pastor the people. You can't just preach to them; you've got to pastor them. You must visit the sick and shut-ins and take them Communion (The Lord's Supper). You must visit them wherever they may be: the hospital, their homes, and the jails. Baptist preachers can get by with being a good preacher and let the Deacons do the visitation of the members, but Methodist preachers must pastor the people." I made visitation one of my highest priorities.

As word got around that I was visiting members, church attendance dramatically improved. I made a point of remembering the names of members and visitors. If a visitor attended church more than once, I remembered their names on their second visit! People really appreciated that personal touch. With the increase in attendance and participation, our financial offerings also increased. The offerings, which had been about $500.00 - $700.00 per week increased to more than $2000.00 per week, and the membership increased to approximately 350 people by the time I left, five years later.

Ikie was a great pastor's wife and mother to our baby girl, Angela (Angie) who was born in 1971. Ikie stopped working outside of our home to become a full-time mother and housewife. She supplemented our income by taking in sewing.

I was tremendously blessed to have well qualified ministers within the congregation who supported me and were very helpful to the success of my ministry: Reverend William Potter (deceased), a former pastor of the church and the Presiding Elder of the Atlanta District Conference; The Reverend Doctor George B. Thomas (deceased), a professor at Interdenominational Theological Center (ITC) in Atlanta; and The Reverend Robert Clayton (deceased), an educator. These three men gave me great support and encouragement. God blessed our ministry in many ways. "To Him be glory and praise!"

It was during my fourth year at Shaw Temple that I was invited to a convention with Campus Crusade For Christ. It was held in Biloxi, Mississippi. While there, I contacted a former seminary classmate, Robert (Bob) Jemerson who was a United States Air Force Chaplain, stationed at Keesler Air Force Base, Biloxi. While showing me around the base I had the opportunity to see what his ministry entailed. One of the highlights was seeing his coffee house ministry, designed for young airmen. The coffee house was a place for airmen to relax and socialize. It was an alternative to bars and nightclubs. I was thoroughly impressed with what I saw with his ministry with young African American airmen. The ratio of African American chaplains to African American troops was extremely low. I saw a great need for our young men to have someone who they could comfortably confide in and relate to. I began to see the chaplaincy in a different light. Up to that time I had not considered the military chaplaincy as a valid ministry.

Bob suggested that if I was not interested in entering the Air Force on active duty, I might consider joining the Air Force Reserve Program. This was a program which would allow me to report monthly for duty. I could still pastor my church and still be available each month to counsel troops and lead worship each month. In addition to the weekend training, you would be scheduled for two weeks of summer training. I came back home with newfound knowledge about ministry in the Air Force. Within days I made an application for the Air Force Reserve Program and was accepted. I was assigned to Dobbins Air Force Base, located about 15 miles in Marietta, Georgia.

The more I was exposed to the Air Force chaplaincy, the more intrigued I became. In the meantime, Shaw Temple was thriving. I had seen the church grow and watched young families gravitate to the church. The average age of the membership was mid-forties and was composed of professionals with college degrees and above. Bishop Clinton Coleman (deceased) had described Shaw Temple as one of the most unique and thriving churches within the denomination.

I found myself torn between the church and the Air Force chaplaincy. My experience with the chaplaincy revealed a great need for African American Chaplains who could better relate to young African American airmen. I prayed about my dilemma and was led to apply for active-duty chaplaincy, was accepted and was assigned to Keesler Air Force Base. I made the announcement of my decision to the church and informed my Bishop. He was shocked and could not understand why I wanted to leave the best and most successful church in his conference. He told me there were eleven other bishops and if I wanted a larger congregation he could assist in that arrangement.

I responded to him by saying, "Bishop, it's not about the money or the size of the congregation. It is about my belief that I have accomplished what I believe the Lord called me to do here. I have prayed about it, and I feel led to enter a new phase of ministry to the military."

He thanked me for my ministry at Shaw Temple and gave me his blessings. I had another challenge that God would help me work through. Just months before my decision had been made, Ikie had applied to Emory University and was accepted in a pilot program. It was the first Master's Degree program in Community Health. Twelve students nationwide would make the cut. Two were Black and two were women. Ikie met two of the categories. She had worked extremely hard to get accepted in the program. It was like writing a dissertation.

Her goal was to finish the Master's Degree, then pursue a doctorate in Community Health. Eventually she hoped to establish a Community Health Center in an impoverished area of Atlanta.

In debating the pros and cons of our futures, she called her mother for advice. Her mother did not tell her what to do, however she said that, "It's not good for a family to be separated." So, Ikie agreed to give up her dream to follow my dream. Nevertheless, she said to me, "I'm going with you, but you better do good and make it worth me giving up what I wanted!"

Chapter 8

FIRST EXPERIENCES ON ACTIVE DUTY

Early years (1975–1978)

"Therefore take up the whole armor of God, that you may be able to withstand in the evil day, and having done all, to stand firm."

EPHESIANS 6:13 (KJV)

I had mixed emotions as I traveled with my family to Keesler Air Force Base in Biloxi, Mississippi. Many thoughts flooded my mind. Major changes always brought on questions such as, "Am I making the right decisions? What if I fail at this? What are you going to do if things don't work out?" My anxiety was based on the knowledge that I had given up a successful career for one that was uncertain. These doubts were produced by Satan. I was reminded that I had prayed about this decision and had asked for the prayers of others whom I trusted. I had felt God's peace and sensed that He had led me to this new venture of ministry.

Arrangements for us at the base had been made. We were met by my sponsor and assigned chaplain, who made sure that we were comfortable. We were given temporary housing until permanent housing was available.

Our children, Angie & Tony

The next day I reported to the base chapel to meet the chapel staff which included 27 chaplains and approximately 20 enlisted support personnel. This was the second largest chapel staff in the US Air Force. There were many clergy from different denominations: Baptist, Methodist, Pentecostals, Lutherans, Roman Catholic, Jewish, Assembly of God and the list went on. The reason Keesler was so large was due to its mission as a training base. It was a part of Air Training Command (ATC).

Getting settled in a military community was quite different from a civilian community. I had to ensure that my immunizations were up to date at the medical Center, get my pay set up in the Consolidated Base Personnel Office (CBPO) and get my name on the housing list for base housing. We stayed in temporary housing for ten days before we were assigned permanent quarters on base. The house was a three-bedroom, with two bathrooms. The neighbors were cordial. One day a neighbor came to meet me and informed us that she had never seen a Black chaplain before.

I had to meet with my immediate supervisor who was a Lieutenant Colonel. We talked about how to get acclimated to this new environment as a military chaplain. One of his priorities was the concept of the "ministry of presence." In this ministry, the focus was on the chaplain being visible to troops. The concept was not completely foreign because as a Methodist preacher that is what I did to a limited degree at Shaw Temple. Here and throughout the military, the chaplain was responsible for visitation to assigned squadrons or units all over the base. We were instructed to take the ministry to the people rather than sitting in our offices, waiting for them to come see us. Chaplains were assigned on a rotational basis to be available for counseling, whether walk-ins or by appointment. This was very important to the commanders because they wanted their airmen to be in a state of readiness at all times. We talked about my job description and some of the requirements and expectations. I was assigned several units for visitation on a regular basis. Being the only African American chaplain on staff was lonely business, because I did not relate to anyone else on staff.

The staff members were appropriately cordial, but not necessarily friendly or helpful. I had grown up in an African American culture, surrounded by people to whom I could share anxieties and express uncertainties. The day I was introduced to the staff I felt like "an outsider." I sensed that they were all indifferent to my newness in the Air Force, strictly focusing on their own careers and personal interests. I got the impression that their feeling was, "You are on your own, sink or swim." I realized that the way I felt then, was in great contrast to what I had felt in my previous work environments. In Canton, I had my family, good friends, and coworkers to support me and give advice on various issues. In Chicago, I had Bill who assisted in pointing me toward the Cross resulting in my salvation. In Richmond, I had Dr. Hart, who provided wisdom and encouragement. These individuals were vital in helping me make the adjustments necessary to be successful in my work. With these professionals I did not feel the same type of support. Ikie had always been there to support me in every venture, even though reluctantly at times. However at Keesler, her hands were full because, not only was she learning to cope with this military life, she also was the primary parent to our two small children. During this same period, our son, Walter Anthony (Tony) was born.

I already knew that my primary job was to conduct the Gospel Worship Service (The Soul Service). This service was predominately African American, which was composed of individuals, both military and retired civilians from the Protestant worship tradition. The civilians brought stability to both the Soul Choir and the Soul Service, because they were stable. Unfortunately, African American chaplains coming on active duty were highly likely to be pigeonholed into leading the service unless they were Roman Catholic Priests. This was natural to some degree and expected, but they were limited to that service only! If you were African American, regardless to your denominational affiliation, it was assumed that you would be comfortable leading the Gospel Service. In many cases this was a false assumption. For example, if you were an African American Lutheran chaplain, a Gospel Service would most likely be foreign to your worship tradition.

African American chaplains were seldom if ever considered to lead a General Protestant (predominately White) worship service. At Keesler, this never happened! I cannot say why this was the case. As I reflect on it now, I believe the chapel leadership felt African American chaplains would not be able to relate to White parishioners. I found that to be a fallacy, years later, when I became the Wing Chaplain at Keesler. I would amplify this fact in the chapter on "Senior Level "in the Air Force. I had been at Keesler for three years and was leaving in one month before I was invited to preach in a General Protestant Worship Service. When I finished preaching that day, the Base Commander's wife asked me if I was new to the base because she had not had the opportunity to hear me before. She had not attended the Soul Service and I had not preached in the General Protestant Worship Service before.

At the end of my first year at Keesler, I received my first report card or Officer Efficiency Report (OER). When I came on active duty the Vietnam War was ending. During the war there had been a huge increase in military personnel. Now that the war was ending the military no longer needed the large numbers, so there was a draw down. This meant that OER's were under "controls." The purpose of controls was to decrease the size of military personnel. For the reader's benefit I should explain what controls meant. OER's would have to be signed off with ratings (i.e., 1, 2, or 3) at three different levels. The levels were:

chapel—my boss (first), Wing Commander (second), and Headquarters Command Chaplain (third). The stratification of the grades progressed from (1) as the highest to (3) as the lowest rating. My boss could give all his captains the grade of (1) if desired. However, the recommendation for the commander would be stratified. At this level only a small percentage would be awarded (1) and (2) the rest would be (3).

At Headquarters Command, the same stratification would take place. So, the percentage of (1) and (2) ratings would decrease even more. For a person to receive a rating of (3) at all levels was a kiss of death. This being the case, the chaplain would have to leave the Air Force.

I had worked extremely hard and expected a report card commensurate with my work. When my OER was completed, I received a (1) at chapel, a (2) at Wing and (3) at Command! When I received my report card, I was very upset. I had never experienced low grades in anything I had ever done. I made an appointment to see the Wing Commander to determine why he felt I deserved a (2) rather than a (1). He explained the process to me, and I discovered that I had not followed the chain of command. I should have met with my boss first. This information reminded me that I did not understand military protocol. When my boss found out that I had jumped the chain of command he called me to his office. He explained that the senior captains who were meeting the Promotion Board to major needed the better grades of 1 and 2's in order to have a chance of being promoted. The thinking was that junior captains such as myself had about six years to recover before the Promotion Board. To console me he said, "Aw, Walt, you'll make major."

This was not a consolation to me for two reasons. First, I did not know if he was implying that I would only reach the rank of major. Second, I had no assurance that the system would treat me fairly. Some of us had no military background and were not provided helpful information related to our careers or upward mobility. I began to distrust the system. In hindsight, I realize that inmates who were in charge were already at work designing my future. By the time I left Keesler as a junior captain, I knew in order for me to succeed, I would have to be observant and shrewd.

Chapter 9

A NEW COUNTRY — ANOTHER BEGINNING

(1978–1981)

"I will instruct you and teach you in the way you should go; I will guide you with my eye"

PSALM 32:8 (NKJV)

I left Keesler and was assigned to my first overseas duty at Royal Air Force (RAF), Lakenheath Air Force Base in England. By this time I was a bit wiser about my career path. After three years at Keesler, I learned a few things that gave me better insights about my future. My eyes were opened to some of the pitfalls I needed to avoid going forward.

I left a training base for a base with a fighter wing, the 51st Combat Support Group. RAF Lakenheath was a base within the United States Air Force In Europe or (USAFE). Everything there was simply different. The climate was different. The people were different and usually friendly. The mission was different and fast paced, with the thundering noise of F-16 fighter planes taking off and landing, sometimes for hours! I was excited with this new overseas assignment. As I reflected on my days and years at Keesler, I had determined that I would seek new ways for excelling in my career.

As stated before, the climate was different, it rained almost daily during most of the year and was cold with snow during winter. The people were different though friendly and very trustful of strangers. Their use of some words that we use had different meanings. For example. A "boot" to us is a type of shoe that is worn to protect our feet from rain or mud. "Boot" to them meant the trunk of a car. Dialects seemed to identify three general classes of people, the Queen's dialect, the educated/middle class dialect, and the common people

When Ikie and the children arrived at Lakenheath, we had to live in temporary quarters until permanent housing was available on the base. We lived for a brief period off base in a local village. Ikie needed to learn how to drive on the "other" side of the car and road the first week she and the children arrived because I had Temporary Duty Yonder (TDY) for a week. It was a frighting experience. I had instructed her to "just think left" whenever she drove.

Lakenheath was a Fighter Wing. The chapel, the hospital, Judge Advocate General (JAG), Consolidated Base Personnel Office (CBPO) etc. supported the Wing (51st Combat Support Group), in the United States Air Force. My job as a chaplain was to support the personnel and families of the Wing.

I found an opportunity to use my skills in teaching a course for parents, Parent Effectiveness Training (PET). I was trained to teach this course at Kessler and was the only chaplain on staff trained to teach this course. I held sessions on base for six weeks at a time. There was a big demand for theses sessions among parents who had children.

It was very helpful and meaningful to parents in learning new skills for disciplining their children. In addition to instructing parents in (PET), I volunteered to provide the Ministry of Presence to personnel at RAF Sculthorpe, a small base attached to RAF Lakenheath, located approximately fifty miles away. I would drive up for visitation of troops stationed there during the week, then drive back on Sunday mornings to provide a worship service.

During the winter months, with black ice on very narrow roads, I traveled there with Ikie and our two small children. Sometimes the roads were simply treacherous. I had purchased a small British car which resembled a VW Beetle, called a Morris Minor. It was dependable transportation and easy to work on. No other chaplain on staff ever offered to help with that ministry. It was tough, but very rewarding. The people really appreciated the sacrifice we made to be available to them and provide ministry.

One of the major benefits of the assignment was being able to travel to many parts of Europe. We were somewhat centrally located and could travel by land or air. We took advantage of travel benefits and were able to visit many places we might had only dreamed about.

The only major downfall of the assignment came when I departed the base after three years. Just before I was due to depart, the base had a major inspection, which included the base chapel. For various and sundry reasons, the chapel failed the inspection. The chapel funds and how they were used along with other aspects of the program were not up to par. Because of my hard work, my immediate supervisor had written me up for a Meritorious Service Medal, which had to be signed off by the Installation Staff Chaplain/06. He had only been assigned to the chapel 90 days, and refused to approve my medal. I left the base having not received any medal at all, a killer for one's career! This was another sign that "Inmates" were in charge.

When I took the training in (PET) at Keesler, I was taking advantage of an opportunity I believed would be helpful for my career advancement later. I was willing to prepare and teach these classes weekly because I was aware of the challenges that parenting brings to young parents. I taught three sessions per year.

Leadership seemed to feel it was necessary to deny me the medal I had worked so hard to earn. I was the only African American chaplain on the base. I had done more work than most of my fellow chaplains. I had exposed my family to dangers on roadways and severe weather conditions going back and forth to Sculthorpe, weekly. I had provided for the spiritual needs of that community, and they assured me that they had been blessed through my sermons, prayers, and visitations. I had only the satisfaction of serving to show for my efforts. I left Lakenheath feeling that I was taking what I felt was "the brunt of the punishment" for the chapel's failure to pass the inspection.

I reminded myself that God had called me into the ministry. He had led me into the military. I was painfully aware that I was at a low ebb in my life. In these last two assignments, my skills and abilities had not been acknowledged by Leadership. I could not show anything for the work I had done or for the sacrifices I had made. The people in charge were calling the shots and I could not do anything about it. I prayed for direction and encouragement. I remembered a message of Jesus in the gospel of Mark, *"And Jesus looking upon them saith, With men it is impossible, but not with God: for with God all things are possible"* (Mark 10:27 KJV) I had to trust that God was still in control and continue to expect His unending faithfulness.

Chapter 10

GOD'S FAITHFULNESS

(1981–1986)

*"Let us hold tightly without wavering to the hope we affirm, for God
can be trusted to keep His promise"*
HEBREWS 10:23 (NLT)

My next assignment came in the summer of 1981. I was assigned to
Air University, located on Maxwell Air Force Base in Montgomery,
Alabama. I came to Maxwell at a very low ebb in my life. I felt deflated,
void, and empty. I wondered what my future would look like. During a
moment of reflection, I thought to myself, *Either my career will continue
spiraling downhill or something will happen to turn it around. If the worst occurs
and I am not successful in the military, I can get out and return to pastoring.*

The thought of getting out of the Air Force was not a pleasant one.
My family and I had given up a lot to be in it. I was aware that I would
be considered for a promotion from captain to the rank of major during
my assignment at Maxwell. At this point, things did not look good for

that promotion. The "controls" at Keesler had prevented me from getting a "good" report card, and because the base at Lakenheath failed the inspection, I had left with a mediocre report and no medal.

My first job at Maxwell was to serve as chaplain to the Squadron Officer School (SOS). This was an in-residence school for young promising 1st Lieutenants and Captains. It prepared them to assume increased responsibilities in their respective career fields. Each group of students would attend class sessions for eight weeks. I also had responsibilities and duties at the base chapel. My office was located there, meaning I participated in all their activities except their academic studies.

I was available for counseling, as needed, by faculty and students. One of the important duties of my job was the involvement with the opening day ceremony to welcome each new class. I delivered the invocation and benediction for these events. I also conducted classes for the spouses who had accompanied the students. These classes were vital in preparing these individuals for their futures as their spouses progressed in rank.

The thought of not moving up in rank did cross my mind at times. However, I had a sense of peace in knowing that God's grace, regardless of the outcome, would be sufficient. My Promotion Board for major met one year earlier than expected, because it was small and was combined with the current year group. "Divine Providence" prevailed, and I was selected for the rank of major. There was joy, joy, joy in the Beamon household! My selection was like being selected "Below The Zone" (BTZ), one year earlier than normal progression. This reminded me of God's faithfulness. No matter how bad things appear, He guides and controls our destiny.

When I arrived at Maxwell, the senior chaplain, colonel/06, my boss, was the laughingstock of the base. For several reasons he did not have the respect of senior leadership. It was well known that he would make inappropriate statements at inappropriate times. Thankfully, he retired during my second year there.

After he retired, my new boss, Chaplain, Colonel Harry Houseman arrived. He was 6'4" with a commanding presence. When he entered a room, there was absolutely no doubt that he was a colonel. His physical stature and facial expressions demanded attention and respect. He was a White United Methodist chaplain, about the age of 50. At his first staff meeting, his demeanor and direct approach left no doubt that he was in charge. He made it known to our staff what his expectations were of us and what we could expect from him, and he did not mince words. Even though I was very apprehensive, he impressed me as being fair and firm.

We had two worship services each week at the base chapel (8 a.m. and 11 a.m.). There were enough chaplains to have two chaplains paired up for each service. The 8 a.m. service was a contemporary worship service. The 11 a.m. worship service was a traditional service, with hymns and anthems. Shortly after Chaplain Houseman arrived, he rearranged the pairing of the chaplains for these services, and chose me to serve with him. I was extremely reluctant to serve him. He was the boss, a "COLONEL," and I had just been selected for major. I would be under his direct scrutiny every week. I tried pushing back for several reasons. First, I felt intimidated by him. Second, this service was the major General Protestant Worship, and the ranking leadership who attended Protestant worship, attended that service. Third, I felt that the Lieutenant Colonels should be in the spotlight, instead of me. He would not budge, so we were paired.

We alternated between leading worship and preaching. He was a stickler for preaching from the Lectionary, in which I had little experience. The Lectionary was a list of scriptures selected for the

Christian year from which we preached every week. I felt I was then limited in my creativity. I could not preach sermons that were on my heart. Even though we both were from the Methodist tradition, my tradition seldom used the Lectionary. Preaching from the Lectionary required discipline; however Chaplain Houseman was the boss and I saluted smartly and used the Lectionary.

We established our routine in worship, and I became more comfortable preaching from the Lectionary. There was a positive aspect from using the Lectionary. The context and content of my sermons became more balanced. But our relationship remained distant. However, one Sunday, after my sermon, something apparently transpired with Chaplain Houseman. Weeks later it was apparent that our relationship had changed. It was never the same again. He took me under his wings, from that point on he gave me opportunities to grow and excel. What initially was a test to serve with him became a blessing in disguise.

Some weeks after our new relationship, one of our chaplains suddenly died. He was the ranking chaplain and there was also a Catholic Priest at Gunter Air Force Station, which was a small base attached to Maxwell located across town. By this time, I had been promoted to major and had served on Maxwell for two years. Chaplain Houseman gave me the opportunity to succeed by selecting me to replace the deceased chaplain. I enjoyed the challenges of this position at Gunter. It was refreshing to be the "boss." I conducted Protestant worship each week and served in many capacities to the Gunter community. I had a great rapport with the people.

I had been at Gunter for about six months when one day my phone rang. On the other end was Chaplain Houseman. My antenna went up as I wondered what the intent of the call could be. He immediately asked me, "Walt, how would you like to go to Air Command and Staff

College in Residence (ACSC). Before you answer, don't even think of saying no." It took a moment for my brain to process the question and his answer to the question. I was shocked and wasn't about to say no. I knew that to be selected to attend ACSC meant that you were considered to be in the top 10% of your peers! That also meant, being handpicked for ACSC in residence was my ticket to the rank of Lieutenant Colonel. Of course I said yes. It was a no brainer. I would not have to relocate my family. When I finished my work at Gunter, he gave me a copy of my OER with a note on the back. The note said, "this will get you promoted to Lieutenant Colonel."

While stationed at Maxwell, after Chaplain Houseman and I developed a good relationship, he shared information with me that I never forgot. What he shared with me helped me understand that all senior leadership are not "inmates," and should not be painted with the same brush.

He told me about Chaplain Colonel Simon Scott (deceased), an African American, who had been had been Command Chaplain for three major commands in the Air Force: Strategic Air Command (SAC), Tactical Air Command (TAC) and United States Air Force in Europe (USAFE). Chaplain Scott had been instrumental in getting Chaplain Houseman promoted to 06 after being passed over in the primary zone. This means that Chaplain Houseman was not selected when he was first considered for colonel. Statistics show that once you are non-selected when considered in the primary zone, the chances of selection in the secondary zone are about five percent. Chaplain Scott helped Chaplain Houseman reach the rank of colonel.

Chaplain Scott had achieved more than any other chaplain, African American or White by serving as Command Chaplain at three of the most important commands in the Air Force, but he never reached the status of general officer. As I pondered this information about the circumstances related to Chaplain Scott's career, I decided that they were "unjust and unfair."

I wondered how a person could accomplish so much and yet be denied promotion to general officer rank! He never shattered the glass ceiling. I am compelled to believe that the reason he didn't was because the "inmates" were in charge.

Chapter 11

THE LAND OF ALMOST RIGHT

(1986–1988)

"And we know that in all things God works for the good of those who
love Him, who have been called according to His purpose."
ROMANS 8:28 (NIV)

I successfully completed the requirements at ACSC and left for my next assignment in 1986. At this point in my career I needed what is referred to as a short tour of duty. It was usually a 1-year remote tour or a tour where your family members could not accompany you. I was given several options from which to choose. One option was to take a tour of Osan Air Base, South Korea, and take my family for two years and meet the requirement of a short tour. It was a no-brainer for me. I felt I needed to be with my family at this particular time. Angie was in high school, and Tony was a pre-teen. The only stipulation in accepting that option was, I had to travel to Korea 30 days before my family could join me.

The trip to Korea was a tremendously long flight. It took 18 hours from the West Coast to touchdown in Seoul, Korea. Upon reaching Seoul, I was exhausted. I had traveled through seven time zones. My body was sore, and I felt like a "board." I wanted to sleep but was told that was the worst thing to do. My body had to adjust to the "day and night" schedule there.

Landing in Korea was like traveling back in time twenty-five years or more. The country was crowded. Rice paddies surrounded the roadways. People worked in groups of ten or more setting out rice plants. The air quality was saturated with the smell of garlic, peppers, and onions everywhere I went. I saw nothing familiar. The homes I saw were small buildings and were occupied by large numbers of people. There were no fast foods like Burger King or McDonalds. The roads were narrow and treacherous. Language was a huge barrier. I found myself making gestures to communicate. A few people could speak some English and they served as interpreters. It was called "the land of almost right" because most things were not perfected.

For example, steps to a building that should have been six inches in height would inevitably have one or two steps that was eight or nine inches in height. If you ordered a plaque, generally you could expect at least one misspelled word. In other words, most things the Koreans did were "almost right."

The chapel staff was composed of two Catholic priests and four Protestant chaplains and four enlisted chapel support personnel. I was the senior Protestant chaplain. My boss, a colonel, was the Installation Staff chaplain. He was a very "unique" individual in several ways. He impressed me as a person who needed to have the center of attention. He did not find it necessary to apologize for abrasive behaviors. He had a habit of intimidating the staff by pointing to the eagles on the epaulets of his uniform. He would literally say to us, his staff, "You see these eagles?" We knew he was telling us, "I am the boss, period."

It was December 1987 at Osan Air Base, South Korea. The atmosphere was buzzing as the annual Christmas party for the Air Base Group swung into full swing. Most attendees wore the appropriate dress for the occasion. My boss, a White, full colonel, defied convention. He appeared in a red suit, complete with a garish green bow tie that pulsated with an array of flashing colors. At an event where decorum and seriousness were expected, he had become the unwitting centerpiece of absurdity, the clown of the night. His antics were so outrageous that his staff discreetly distanced themselves, unwilling to be associated with his spectacle. Little did I know that this spectacle would be overshadowed by a revelation that would change the course of my career.

Shortly after that unforgettable evening, the results of the Promotion Board for the rank of Lieutenant Colonel (05) were released and to my astonishment, my name was on the list. It was a moment of personal triumph, a testament to years of dedication and service. However, the spectacle of my boss' antics at the Christmas party would soon cast a long shadow over my accomplishments.

Our Protestant Parish Council hosted a monthly luncheon for the Protestant chapel community At the next luncheon, when it came time for the senior chaplain, my boss was asked to give remarks. As he announced the names of chapel members who had been selected for promotion, I awaited my moment of recognition with anticipation.

What followed, however, was a shocking declaration that would reverberate through the room. When he reached my name, he made an utterance that left me bewildered- a reference to "Peter Principle." The words carried an undertone of disapproval, and although I couldn't grasp their full meaning at the time, I knew they were far from complimentary. At that, he had effectively branded me as an individual who had risen to a rank beyond their competence. It was a crushing blow to my self-esteem and professional pride.

What made the situation even more bewildering was the fact that I had achieved this promotion in my 12th year for Lieutenant Colonel, when statistics showed that the average length of time is 15 to 17 years! I had exceeded expectations, yet my boss, the very clown of the base who had turned a dignified gathering into a circus had labeled me as a symbol of ineptitude.

I was devastated. I felt I needed to talk to someone I trusted to share my feelings. So I picked an Army Chaplain, a colonel with whom I had a good relationship. After I explained what had happened with my boss, he asked me two questions. The first was, "How long have you been in the Air Force?" I responded, 12 years. His second question was, "You have been selected for Lieutenant Colonel? Walt, don't let that incident bother you, he's jealous of you. Just keep doing what you are doing, brother!"

This incident marked a turning point, a stark realization that the chaplain, who had become the laughingstock of the base, was emblematic of a more profound issue—the prevalence of an untrustworthy top-level leadership, lacking vision, and tainted by corruption. This marked the beginning of my journey to expose the cracks in the system and uncover the truth about those inmates in charge.

Chapter 12

MY FIRST RODEO AS A WING CHAPLAIN

Whiteman Air Force Base
Knobnoster, Missouri (1988–1991)

"Brothers, I do not consider that I have made it my own.
But one thing I do: forgetting what lies behind and straining
forward to what lies ahead, I press on toward the goal of
the prize of the upward call of God in Christ Jesus."
PHILIPPIANS 3:13-14 (ESV)

We left Osan Air Base during the summer of 1988 and headed to Whiteman Air Force Base in Knobnoster, Missouri, located about 65 miles South of Kansas City, Missouri. Whiteman Air Force was a part of the Strategic Air Command (SAC). The mission was to provide protection of the country by Minuteman missiles. (A few years later it became a B2 Wing.) Whiteman was in a community that did not want it to grow. It had a few stores and a couple of fast-food stores and a grocery store. We were told that the community did not want Knobnoster to grow because they did not want the issues associated with urban development. Sedalia Community College was located ten miles from the base. I had been assigned as the Wing chaplain, the senior ranking chaplain. My staff was composed of a Protestant major (the Senior Protestant Chaplain), a Protestant Captain, one Roman Catholic Priest (Major). I had an enlisted staff of three, the ranking member was a Master Sargent. There was also a full-time secretary.

The priest had not been selected for the rank of Lieutenant Colonel after being considered by several Promotion Boards. He became my first challenge at my new job. Naturally, he had been disappointed with the result of the Promotion Boards and lacked the motivation to be an effective staff member who would contribute to my goals for the chapel program. He let me know, on my arrival, that he was only going to perform the most basic duties of his job. This was primarily to provide Mass for the Catholic Community. His position was, "I will not do anything to hurt the work of your chapel program, but don't expect much else from me." I respected his honesty and transparency. I understood the difficulty of functioning effectively in a position that had failed to push one into the next rank. He refused to function on any other projects that we provided to the base community. Therefore, the other chaplains had to take on additional responsibilities.

Another challenge that I faced in my new job as Wing Chaplain was with the base commander. He was a "hot shot" young colonel who was "bucking" to become a general officer. I found it difficult to establish a good working relationship (one based on trust) with him. It appeared that he preferred to interact with the Senior Protestant chaplain rather than with me. Often, he would provide information to the Senior Protestant that should have been provided directly to me as the head of the base chapel. In this dilemma, I had to find ways to navigate around him.

In this position I was exposed to a new level of information. I learned more about the overall operation of the Air Force. I interfaced with commanders from different divisions on the base and became familiar with their specific roles. I had to attend meetings at the Wing and Base commander levels. This helped me understand the major role that each played in the complete operation of the mission.

Also, in this new position, I became keenly aware of issues that African Americans were having that were unjust and unfair on their jobs. Periodically, the Air Force would send us on Temporary Duty (TDY) for training purposes. It was in these settings, during free times such as lunch or other breaks, that we would talk. We would get together and share the struggles we were having on our bases. We discussed issues that were affecting our careers in negative ways. We knew that government phones on our bases were subject to be monitored. Each government phone had a red sticker on it with the words "Your conversations are subject to be monitored." The phones were only to be used for official business. In those days, personal cell phones were limited or nonexistent. So, we had to take the opportunities to communicate during TDYs. As I heard about the concerns of other chaplains, I was reminded of my own experiences. I remembered the turning point in my desire to change the "status quo." I recognized that I was not able to do anything but listen and was distressed. It was distressful because I wanted to help.

Overall, this assignment was a good learning experience. I learned about the operation of the base in achieving its mission, but also about the lack of responsiveness of Air Force Chaplaincy leadership to unjust, unfair issues that African Americans faced. I survived my first challenges as a Wing Chaplain. The skills and knowledge I learned in this job prepared me for bigger challenges that I would face in the future.

Chapter 13

"ANOTHER NEW VENTURE" AIR MOBILITY COMMAND

Senior Level (1991–1992)

*"Be anxious for nothing, but in everything by prayer and supplication,
with thanksgiving, let your requests be made known to God."*
PHILIPPIANS 4:6 (ESV)

In 1991, after three years at Whiteman, I was assigned to command level. The assignment was to Air Mobility Command (AMC) located on Scott Air Force Base, in Belleville, Illinois, just outside St. Louis, Missouri. This assignment was a step toward upward mobility in my career. Potentially this could lead to a higher promotion. It was supposed to be a "controlled" tour. A controlled tour was one in which I could stay in that job on that base for three years before being reassigned. That was very important to me because I had promised our son, Tony, that he would be able to complete high school at O'Fallon High School, in O'Fallon, Illinois, where we were living.

My job title at command was Chief of Plans and Programs. I was responsible for coordinating the needs of various programs such as Vacation Bible School (VBS) and other similar programs for the bases within AMC. The job also entailed planning and hosting an annual conference for Wing Chaplains and chapel support personnel from all the bases within AMC.

During the Spring of 1992, my second year at AMC, my board for selection for colonel met. I had been made aware that Air War College in residence, seminar or correspondence was becoming a requirement for selection. I had not completed this requirement. I surmised that I was unlikely to be selected. To have completed that requirement was no guarantee of promotion, but without it your chances were significantly less.

Another prevailing factor that clouded my optimism about being selected for colonel, was that I had a rating of "3" at the command level for each of the three years of my assignment at Keesler. If they were not removed, they would have a negative impact on how my records were scored when the Promotion Board met. The "3" ratings were the result of mandatory controls and did not accurately reflect my ability or capability to perform my duties. Before my official board met, a friend strongly advised me to contact the command chaplain who had rated me at my first assignment, Keesler, with those "3"s and request him to write letters to the Official Board of Air Force Records. The purpose of the letters would be to have the "3"s removed. If mandatory controls had not been in place, I would have received "1's" instead of "3's." I followed through on his advice. The Command Chaplain removed two of the (3's) and left one.

A third reason that I was apprehensive about the potential outcome of the Promotion Board was based on the relationship that existed between the Chief of Chaplains and me. We had interacted on several occasions, and I did not feel "warm fuzzies" from him. However, I was a military man and knew that "warm fuzzies were not required, nor should be expected to perform my duties. But in this case, I felt his attitude toward me would have some bearing on my selection for colonel. My Promotion Board consisted of chaplains (colonels), line officers (colonels), and a general officer who headed the board.

One of the chaplain colonels on my board was a friend of mine. I was not aware he sat on my board until after the results were published. He indicated that another chaplain colonel had suggested a different African American for selection instead of me. My friend made the assessment that I was a much better chaplain than the person who had been suggested. He had known both of us for many years and knew that my work, in every way, was superior to the other chaplain. He was now faced with a dilemma, He had to decide which of the two he would give the highest score. The chaplain who received the best score from the board would be the selectee. He gave me the highest score.

After the Promotion Board results were released and my name was on the list, my friend called the Chief of Chaplains to find out what his feelings were about the board results. He was told by the Chief, "Chaplain _____ you did what you felt you had to do, now I must do what I feel I have to do." My friend was curious about what the general meant. He was told, "You will find out later." During those days the Air Force was "top heavy" in the rank of colonel. A board had been established to trim the number of colonels (06). It was known as the Selective Early Retirement Board, (SERB). He informed me that within months of my selection, this board met. When the board results came out, his name was on the (SERB) list. He then understood what the Chief of Chaplains had meant from his earlier statements. This meant he was forced to retire within months.

Inmates, in the context of this book, are those in leadership positions in the Air Force who are imprisoned by their mentality and thought process that "African Americans should be governed by their decisions and desires." The scenario I just described demonstrates how the inmates functioned. My friend made a decision that was not acceptable in the Chief of Chaplain's thinking.

Prior to knowing the results of the Promotion Board, I had tried to prepare Ikie for the likelihood that I would not be selected. We both had surmised that to retire as a Lieutenant Colonel was nothing to be ashamed of and that we could live with that reality. After all, I had never set colonel as my goal in the first place. I did want to go as high in rank as I could. However, I wanted to be effective in my work regardless to the rank I achieved. I was acquainted with chaplains who were very rank conscious. I saw some who would kiss up to commanders, the Chief of Chaplains, and anyone they thought would put them in a position to achieve a higher rank.

When the results of the Promotion Board were released, I was speechless when my name appeared on the list. I praised God repeatedly for His grace, that unmerited favor! My line number on the list was near the very end, which meant I would have to wait 18 months for promotion or pinning on the eagles; but there was an expression that went something like this; "happiness is having a line number!"

So now, all bets were off as it related to when my next assignment would occur. When I came to Scott Air Force Base it was a control tour of duty (Three years). With my selection, I was eligible for a new assignment. Everything I had promised Tony about his graduation from O'Fallon High was in jeopardy.

A few weeks after receiving word of my selection, the personnel division chaplain in our office approached me with a situation pertaining to a possible new assignment. He indicated that the Chief of Chaplains, had selected a chaplain (colonel), who was the Commandant of the Chaplain School. The Commandant had been nominated to become the Wing Chaplain at Keesler Air Force Base, Biloxi. The Commandant is the senior ranking person at the school. The 2-star general at Keesler, at 2nd Air Force, had rejected the nomination. Instead, he requested that the Chief of Chaplain send him three names and he would pick the incoming Wing Chaplain.

To me, that was a red flag as it related to the Commanders' confidence in the Chief of Chaplain's judgment! My name would be one of those submitted. The personnel chaplain in our office, assured me that with me being a brand new 06 selectee, I had nothing to worry about. Somehow, I was not convinced by his logic. I did not want this assignment. There were several reasons for my apprehension. Namely, it would uproot my family at a critical time in my son's life. I also knew that Keesler had the reputation for being a "dumping ground" for chaplains, some of whom were less desirable for better assignments, or were near the end of their careers, just waiting for their retirement. I knew that to improve the morale there would require changing how business was done, which would not be an easy task. Keesler was my first assignment, and I knew what the environment was like.

Three weeks after the personnel chaplain had approached me about the job at Keesler, he came back to announce that I had been selected. I simply could not believe the news. Although, I could never fully buy his rationale of not "having anything to worry about" I was still in shock of the revelation that I had been selected for that job. I had one of two choices in this matter; put in my papers to leave the Air Force or accept the assignment. I accepted the assignment, so Ikie and I started making plans to relocate to Mississippi.

We decided to take a trip to the area to look for a house. When we arrived in Biloxi, we decided to go by the base to say hello to the secretary, whom I had known since my first assignment there in 1975. While we exchanged some pleasantries, reminiscing about the days gone by, one of the senior ranking chaplains, a lieutenant colonel, came by. He introduced himself and we talked briefly. The senior enlisted non-commissioned officer also came by, and we talked briefly.

Upon our departure from the chapel, Ikie, who has the spiritual gift of discernment, had some uncanny feelings about both guys. Her words were, "You are going to have trouble out of the chaplain and the enlisted guy is gay!" I tried dismissing what she had told me by responding, "Well, we know that some guys have effeminate traits, however that doesn't necessarily mean they are gay and you could be wrong about the chaplain giving me problems." She responded by saying, "Okay, I hope you are right."

Chapter 14

CONFRONTATION— DECISIVE LEADERSHIP

Keesler Air Force Base

Senior years, (1993–1997)

"Count it all joy, my brothers, when you meet trials of various kinds, for you know that the testing of your faith produces steadfastness."

JAMES 1:2 (ESV)

Fast forward the time to August of 1993. We had now moved to Keesler. It took a few weeks to get situated in our new home. I met my staff of twenty chaplains, which included three Catholic Priests whose ranks were as following; a lieutenant colonel, a major and a captain. The Protestants ranks were three lieutenant colonels, three majors, eight captains, two 1st lieutenant, and one 2nd lieutenant, plus three Protestant Reserve chaplains, who were captains. There were ten enlisted members with the ranking member in the grade of Senior Master Sargent.

Keesler had the distinction of having the second largest staff of chaplains in the Air Force. God, in His Divine wisdom, had seen fit to put me in this place at this time of my career. My prayer was that He would prepare me for the challenges ahead and use me to His glory for the good of the staff and the base community. Being selected for colonel and becoming the Wing Chaplain was not the most important thing to me but using the opportunity to improve how the chapel functioned for the greater good was most important.

In my first meeting I laid out my plans in terms of what I expected of the staff and what they could expect from me as the new Wing chaplain. We had a cohesive, congenial meeting. The staff seemed supportive. However, I sensed that the morale of the staff was low. I was told by the staff that this was due to the behavior of the previous Wing Chaplain who in their opinion was, "intimidating, like a dictator."

My next step was to meet my boss. She was a one star, brigadier general. As the Wing Commander, she made me aware of her priorities for the Wing and her expectations of me. My marching orders were to go back and align the chapel priorities with the Wing priorities. The entire base had marching orders from Air Education Training Command (AETC) Headquarters, which were in line with Air Force priorities. The whole base developed metrics which would ensure that we were working smarter and more effectively, thus saving dollars in the process. We put together a team comprised of chaplains and enlisted members which met weekly to develop metrics which would allow the chapel programs to dovetail into the Wing's mission.

After three weeks of my arrival, I joined the Saturday morning group of golfers, which were officers of the base. I had learned to play golf late in life at the age of 48. On this particular Saturday morning, the 2-star, major general of 2nd Air Force approached me and indicated that he wanted the two of us to play together that day. I knew that something was up. I just could not figure out what might be on his mind. He was very cordial and expressed appreciation for my being a member of team Keesler. We had played about three holes when he changed the subject and asked me point blank; "Walt, do you know the story of how you got here?" I responded by saying, "Well sir, I'm told," and he cut me off before I could tell him. "Well, I'm going to tell you how you got here. The Chief of Chaplains wanted to send someone here to retire and I told him, not only no, but hell no! I told him to send me three names and I would select the one I wanted. Your name was one of the three, Walt, I know your golf handicap. I know all about your record. Walt, you have one of the toughest jobs on this base. I look at you as any other commander on my base. And no commander is selected on this base without going through me! That's how you got here, and you have my support!"

This revelation about the Chief of Chaplains, having nominated the wrong person, as perceived by the commander, was another example of how a lack of vision and White privilege at the highest levels of the Air Force chaplaincy existed! This 2-star general, and commander, distrusted the decision of the Chief of Chaplains. He did not believe the Chief of Chaplains had nominated the best leadership for his base! What a tragedy! I recognized that in my position as Wing Chaplain, I had to make decisive, effective, tough decisions in order to improve the functions of the chapel program.

It was during one of the chapel metric team's meetings in 1994 that I had my first confrontation. The metric team was composed of a cross section of chaplains and enlisted chapel support personnel and represented the four divisions of the chapel. The divisions were the: hospital, student, parish, and executive. I proposed making changes by moving people around to ensure the best efficiency and use of resources. My decisions were based upon my conversations with the 2nd Air Force Commander and the Wing Commander.

It had come to my attention that the chief of the student division, a lieutenant colonel, would disappear from his office, would not use the sign-out board as to his whereabouts and would be gone for hours. There was no accountability. Based upon his lack of contributions in staff meetings, task force meetings and lack of communication with other staff members related to our mission, my perceptions were that he was lazy. The student division was located a mile across the base. During this meeting, I announced that he would be moved to the Larcher chapel where my office was located. He openly took offense to being moved, in front of junior chaplains and enlisted staff members. I listened to his rationale for not wanting to move, then stated that my decision was final. He continued *to* openly *display* his opposition by grumbling under his breath. He displayed his anger by turning red, twisting in his seat with raised eyebrows. His appearance said, "How could a new Black Wing chaplain move me like this!" It was apparent that his behavior was being keenly observed by the rest of my staff. At that moment the Holy Spirit reminded me of the words of the 2nd Air Force Commander, "Walt, you have one of the toughest jobs on my base!"

I knew I had a major problem that had to be dealt with right then. I said to him, calmly but firmly, "Chaplain _____ when this meeting is over, I need to see you in my office." Most of the people there had an idea of what was about to happen. I was later told that a young junior chaplain knew that what he had just witnessed was a serious offense and whispered to another chaplain, "Chaplain _____ just pissed in his cheerios!" As an officer, you are taught from day one to never disrespect a senior officer, especially before junior officers and enlisted!

For the lieutenant colonel to argue with me, in front of junior officers and enlisted was the epitome of disrespect! He could have easily said, "Chaplain Beamon, could I speak to you later about the problems I see with your proposal?" I would have welcomed the opportunity for a discussion. There was no way I could allow him to get away with what he did. I cannot say what he might have been thinking or what his motives were. My belief was that he did not want to work with or for an African American. I believed that he felt he could get away with disrespecting me.

After the meeting, he met me in my office. I said to him these words; "Chaplain _____ you are a senior officer, and you know quite well Air Force policy as it relates to disrespect of a senior officer before junior officers and enlisted. As a result of your behavior, I am going to prepare a letter of reprimand and you are no longer the branch chief of the student division!" His response, in an angry tone to me was, "Well, if I go down, I'm taking you with me!" This was a veiled threat to me. My response was, "Do what you please, but you are going down!"

I dismissed him and proceeded to pay a visit to my Wing Commander. I briefed her on what had transpired and my response of preparing a letter of reprimand and relieve him of his duty as the branch chief of the student division. I said to her, "General, from the actions I plan to take against this chaplain, I know there are going to be some fireworks. What I need to know from you is whether you approve of my plans and whether I can depend on you for your support in what I know is coming."

I shall never forget her counsel to me. Her words were, "Walt, that's what the Air Force pay you big bucks for, to make tough decisions! Yes, I have your back, just make sure you keep me in the loop!"

That was all I needed to know. I knew I was working for a commander with integrity. I went back to my office and wrote the letter of reprimand which had to be reviewed by the Judge Advocate Office (JAG) and filed in the personnel office. The chaplain was issued the letter and I informed him that he was relieved of duty as the student division chief. I further informed him that he would be given an office in the chapel where I was located.

He went to the Equal Employment Opportunity and Treatment Office (EEOT). He filed a complaint against me with nine charges. He made trouble by contacting people in the Protestant community soliciting their help in defending him. When he filed his complaint with EEOT his case was forwarded to Higher Headquarters of AETC for investigation.

Ikie, fearful of the ramifications of my actions, pleaded with me not to pursue corrective actions. I personally agonized over it, but three things were very clear to me as I wrestled with my dilemma. First, I believed deeply that I had made the right decision in reprimanding this chaplain for his blatant disrespect. I also believed that he was lazy and not worthy of promotion to a higher grade. Second, I knew that if he were allowed to get away disrespecting me as his boss, I would lose credibility and respect with the rest of my staff. Third, I knew that win, lose, or draw, with the pending investigation from AETC, I would pay a price! For me, it was always about principle, right versus wrong. I reasoned that regardless of the outcome I would be able to look at myself in the mirror and like what I saw. Win, lose or draw, God had blessed me beyond anything I could have ever imagined. I was determined to do the right thing for the right reasons. Was I fearful in this situation, not knowing what the outcome would be? You bet I was! I knew that I had no support from the Command Chaplain's office at AETC, nor the Air Staff Chaplain's office.

I was still in the process of aligning the mission of the chapel with the mission of the Wing. I made the decision to shake up the Protestant side of the house. The chapel services were still set up as they were during my first assignment in 1975. That is, White chaplains conducted General Protestant services and African Americans were still pigeonholed in the Soul Service. I made what some would call radical changes. I moved a young gifted, African American chaplain from the student division to the predominately White general Protestant service as the leader. I knew he would do a superb job. I spoke to the African American chaplain, who was the leader of the Soul Service about sharing it with a White chaplain. They would work as a team. He disagreed with me and indicated that he did not believe it would work. He believed that he would overshadow the White chaplain. His position could be seen as reversed racism. I made the decision to relieve the African American chaplain from his serving in the Soul Service. Therefore, the White chaplain had full responsibility for its leadership. He and his wife did a phenomenal job with the service!

There was excitement in worship. They loved the congregation, and the congregation loved them, because they were authentic. Whenever the Soul Choir was invited to sing in the community, the chaplain and his wife were there to support them.

After relieving the African American chaplain from his duty, he was upset and decided to align himself with the deposed lieutenant colonel, who had been fired from his job. So, they conspired and sent a package of documentations and allegations against me as the Wing Chaplain, addressed to the President of the United States! They didn't know that I had a good friend, Chaplain, Colonel John Blair (deceased), who was stationed at the Chief of Chaplains Office. Their letter never reached the President. At some point along the way it was determined by someone in the chain of command that the letter should not go to the

President but was forwarded to the Chief of Chaplain's Office for his attention. When it reached that office, Chaplain Blair retrieved it, called me, and informed me about its content. To this day they never knew what happened to their letter. Bottom line, I had shown equal justice in my decision to fire both a White and African American chaplain. There was no discrimination in my actions. There was one bright spot at this time. Eighteen months had expired from the time I had been selected for colonel and now the time had finally arrived for promotion and pin-on. I needed something good to occur to take my mind off of dealing with issues regarding staff members and disciplinary problems. My chapel staff had invited my family and people across the base to the chapel to join in the celebration. My immediate and extended family members were present for the pin-on celebration. There were approximately 100 people to assemble and wish me well on this joyous occasion. It was especially wonderful to have Ikie and Tony to do the honors of pinning on my new rank of colonel.

Ikie and Tony pinning my rank of colonel

Lajune Buckley, Walter, Ikie, Angie, Cedric Buckley and Reginald Buckley

After a couple of years, I moved an African American chaplain from leadership of the general Protestant worship service to give him more exposure and to help him get promoted. He did an outstanding job and was very likable. He was so well liked by the congregation that when I moved him, a committee of seven white parishioners made an appointment to see me. They were upset that I had moved their chaplain. I felt it was necessary for them to understand the reason for my actions. I explained to them that if they wanted to see their chaplain promoted, it was necessary to move him into various positions to give him a breadth of experiences prior to his meeting a Promotion Board. In other words, the moves I was making could potentially help their chaplain get promoted. They reluctantly accepted my reasoning, and now understood my rationale.

I also understood how they felt. It is difficult to get use to a chaplain or other leader whom you are comfortable with and then that person is moved. In the past, some Wing Chaplains would allow chaplains to remain in the same position for the entire length of their tour. This did not help young chaplains grow and become more capable of serving in other capacities. The base parishioners wanted their chaplain to get opportunities to grow and move up in rank. They were usually happy whether their chaplain was Black, White or polka dot.

I mentioned earlier that I was under investigation by Headquarters at AETC. The team started their investigation in the Fall of 1994. It was headed by a very nice White colonel line officer. The team spent a week gathering information by interviewing chaplains, support staff and parishioners. It was a tense, disruptive period for my administration. The investigation lasted a week, just before the holidays of Thanksgiving through Christmas.

I prayed for God's protection and peace. I knew in my heart that I had taken the right actions with the staff members who were responsible for the investigation. I vowed to wait on God for deliverance. This was difficult, especially during the Holiday Season. I would have to wonder and second guess the findings of the team for almost three months before the report would be released to my Wing Commander, then to me.

The AETC Command Chaplain made a visit to the base while we were waiting for the results to be released. I shall never forget that he and I, along with our staff, went out to dinner one evening. I went to his lodging to pick him up. I already knew that he was not supportive of my actions regarding the chaplains I had removed from duty. While in his room he proceeded to tell me that I couldn't fire a senior chaplain as I had done. That angered me and I stood up, walked over to him, toe to toe, eyeball to eyeball, and said, "Chaplain _____, I did it and will not change anything, without a direct command from my Wing Commander! Chaplain _____, I don't work for you! I work for the Wing Commander, period! End of conversation." He didn't like what he had heard from me and dropped the conversation.

By nature, I am generally a kind person. My mother was a sweet loving woman. My dad was one with quite a bit of patience, until you rubbed him the wrong way, then all bets were off because he showed his anger no matter who you were. I have some of both traits in me. In hindsight, I must admit that a trait from my dad came out that day!

After what felt like an eternity, the report of the investigation was released. It showed that the (9) charges against me were "unsubstantiated!" After the results were released, the (06) leading the investigating team called me. We had an encouraging conversation. He complemented me on the program I had developed, including the cohesiveness of my staff. At the end of our conversation, he said to me. "Chaplain Beamon, I need to tell you this, if you had not been Black, this investigation probably would not have been done!"

My staff was very pleased with the result of the investigation. They approved of my actions and leadership. One of the majors on my staff came to me one day and said to me, "Chaplain Beamon, had you not taken the actions that you did, I know for a fact that you would have lost your staff. They would have lost respect for you."

A few years later around 1996, I was selected to sit on several Promotion Boards. One of the boards was a (06) board. I used whatever influence I had to get a fellow African American chaplain selected, who had been passed over in the primary zone. You must understand that if you have been passed over in the primary zone for (06), your chances for selection the next time around are severely diminished.

In other words, your chances were slim to none. This was a chaplain I had known for more than 20 years. He loved the Air Force and its people and he was a superb chaplain. His Wing Commander was so impressed with his work ethics that he insisted on him deploying to the war with him.

In the scoring of his records, there was a split between my score and the score of the chairman of the panel, a 2-star Air National

Guard General. The scoring is set up on a scale from 1-10, with 10 as the highest rating. I rated the chaplain a 9.5 score. I seem to recall that the general rated him a 7.5. The other raters on the panel rated him between 8 and 9.5. A split of 2 or more points had to be resolved between the two of us. That meant I had to lower my score, or he had to raise his score to resolve the split. He and I discussed how we arrived at our rating for more than ten minutes. I think, for the sake of time, when he saw that I was firm with my score, he made the decision to raise from 7.5 to 8. It turns out that that the .5 was all that was needed for the chaplain's selection to 06.

During this assignment, I had prayerfully confronted the major challenges that I had to face. God gave me wisdom and courage to make tough decisions. I proved that radical changes could be made successfully, and I worked hard to ensure that all the chaplains on my staff and others in the Air Force had opportunities to grow.

Recently I came across a book by Max Lucado, entitled, *Facing Your Giants*. This book focused on the biblical narrative of David and Goliath. One of the powerful statements he made in the book was:

"Focus on Giants—You Stumble
Focus on God—Your Giants Tumble"

In 1997, after four years serving as the Wing Chaplain, making the controversial decisions that I had made, I knew there would be some repercussions. The inmates would not allow me to move on in my career without some form of punishment. I did not know what it would be or how it would happen, but I prepared my heart and mind for "something" to come down. I was also fully aware that it was time for a new assignment. I wondered where that would be?

Chapter 15

INMATES PAYBACK — UNITED STATES SOUTHERN COMMAND

(Senior Years 1997–1999)

"The Lord will fight for you, and you have only to be silent."

EXODUS 14:14 (ESV)

Word reached me that personnel at the Chief of Chaplains Office had determined that I would be offered an assignment that I probably would not like. The intent was to make me seriously consider retirement. I thought about this information and determined that the inmates would not be able to force me from service, because of my actions; but they could make my life more challenging.

A chaplain's conference was held at United States Air Force in Europe (USAFE) in Germany in 1997. One of the major functions of this conference was to discuss and make assignments. My informant, a fellow chaplain, told me that a decision had been made to offer me an assignment to United States Southern Command. This command had been functioning in Panama for years. It was a unified command, is one which personnel from all branches of the military; Army, Navy, Marine and Air Force function as a unit. The Command worked to build regional and inter agency partnerships to ensure the continued stability of the Western Hemisphere and the forward defense of the U.S. Homeland. The Command was in the process of moving from Panama to Miami, Florida.

Accepting this assignment would mean that I would be the Command Chaplain. My duties would include providing the spiritual needs and the ministry of presence to the personnel of the command and beyond. I would have to travel to various countries in South and Central America. During these visits, I would meet with Ambassadors, Commanders, and the troops. I was made aware that there would be no Air Force support personnel to assist me in the performance of my duties. Awareness of having to travel without support and provide for the spiritual needs to Headquarter personnel also without Air Force support was very daunting. This was my punishment. I had two options. I could accept this assignment, or I could get out and serve a congregation as a pastor. Air Force policy said that once an assignment is made one would have seven days to accept it or opt to retire within 12 months.

When my career was not going well, I had considered the possibility of pastoring again so that was not a foreign idea. Also, we were buying our home and Ikie had a full-time job. I decided to call the Bishop over the West Florida - South Mississippi Conference of the AME Zion

Church. I made him aware that I was contemplating retirement. I wondered if the church in Pascagoula, Mississippi was available or would be soon. If so, I would be interested in serving there. His response was, "have you considered starting a new church where you live? "I was simply amazed at this response given I knew that the denomination would not provide the necessary resources for such a venture. Another issue was that I would be on my own in trying to establish a new church without mission support. I was already in my senior years and this venture did not appeal to me. I was greatly disappointed because a few years before, he had said to me, "Call me when you get ready to retire, the church needs you."

When I heard what was in the making at the Air Staff, I did a bit of soul searching, considering the "pros" and "cons" of accepting the assignment. I prayed about the difficult choice I had to make. I remembered an older African American chaplain, whom I had known since 1975, who was still on active duty. He was Chaplain, Colonel I.V. Tolbert (deceased) whose experience and wisdom I respected. I explained the issues I was facing and the decisions I had to make. After listening, he calmly said, "Walt, no matter where they send you, staff, or no staff, they've got to pay you colonel's pay. You can't eat prestige."

I reasoned that I had about five years left before mandatory retirement. So, I was determined that I would not allow anyone to force me to retire. In fact, I vowed to myself that I would stay until the very last day, that I was eligible before mandatory retirement. I knew the intentions of the inmates. They wanted to force me to retire. When the assignment was made, I had seven days to accept or retire. I called the personnel office at the Air Staff and indicated that I would take the assignment. There was complete silence on the other end of the line. I sensed that the chaplain on the other end was surprised by my acceptance of the assignment. I indicated that I would be requesting funds to fly to the Panama Canal Zone to meet with the 4-star commander of US Southern Command and the Command Chaplain. That was arranged and I met with the 4-star commander and his chaplain to get a briefing on the move and more about my duties.

The Command Chaplain was helpful. He gave me several pointers and took me to meet the general. The general was easy to talk to. He provided information about what the Command had been like before and what it may be like after moving to Miami. I was looking forward to working for him. Little did I know, at that time, he would not be coming to Miami but going to Supreme Headquarters Allied Powers Europe (SHAPE), the Headquarters for NATO.

When the family arrived in Miami and found our way to US Southern Command, we found a huge multi-story, newly constructed building which housed the staff for the entire command. It was in a large open field, near Miami International Airport. There was no housing for families nor infrastructure for a military community. Everyone assigned to the command was on their own, living in the Miami area. Housing in Miami was very expensive, and it was very tough for junior officers and enlisted families.

Because there was no military housing, we chose to purchase our home. It was in the Perrine community, about 20 miles from headquarters, where my office was located. I had to travel on a very busy interstate getting back and forth to work. Often there were serious accidents on this thoroughfare. Sometimes it would take an hour to reach work or get home after work. Miami is extremely multicultural. Ikie had to navigate in a predominately Spanish speaking environment. I was reminded that this assignment was indeed my punishment for firing a lazy, disrespectful, White lieutenant colonel chaplain. It did not matter that I also fired an African American chaplain. I had shown no favoritism. Everyone under my supervision had an equal opportunity to succeed. Thats what the inmates in charge never understood!

I found it extremely difficult to provide for the many spiritual and emotional needs of the personnel in the circumstances that I had to work. I provided as much counseling for domestic abuse, overindulgence of alcohol, and depression as I possibly could. Some personnel had referrals to civilian mental health professionals. Due to expense, some personnel would send their families back home and they would stay in the area.

After several months at Southern Command, a Navy Captain (06), and I were in conversation. He asked me about my staff. I indicated that I had no supporting staff, to which his response was, "Chaplain, what do you mean that you have no staff?" I made him aware that the Air Force decided to punish me by sending me there without support. He then said, "Chaplain, we've got to fix that, I'll tell you what, let me see if we can't fix that. How do they expect you alone to provide ministry to us?"

Within two months, after this conversation, I had a Navy Petty Officer, E-5 assigned to me. She was a reservist brought on to active duty. She was a hard worker. She assisted me in my necessary visits to other countries and made sure that we had everything we needed at the headquarters and for the ministry of presence trips. She was a Godsend.

The inmates in charge, meant it for evil by assigning me to a MAJCOM without a single person to support me. I had left a staff of more than 40 personnel at Keesler, including a secretary, to be assigned to a job without any support. They meant it for evil, but God meant it for good! Psalm 37:25 from King David states, *"I have been young, and am old, yet have I never seen the righteous forsaken, nor his seed begging bread."*

These words rang true to me as I struggled with the treatment I received, as "payback" for doing what I believed was the right thing for the right reason. God provided the help I needed.

By the time I got settled in my job at Southern Command I had two years of retainability left before mandatory retirement. I did not want to retire outside of the Air Force.

My good friend, John Blair, was still working at the Chief of Chaplains Office.

He was my primary contact for what was going on in the Air Force, more specifically, the chaplaincy. I called John with the hope that he could use his influence with the Chief of Chaplains to find another assignment in the Air Force. John approached the Chief of Chaplains. According to him, the conversation went like, "Chaplain _____ you need to move Walt from Southern Command and allow him to spend his last two years in the Air Force. Chaplain _____ you need to do the right thing by Walt. You know that you should have assigned him to USAFE instead of sending Chaplain _____ there in the first place." The chief's response was, "Well, I don't know if I can sell Walt to a commander or not." John's response was, "Chaplain, you won't be able to sell him to a commander if you don't nominate him! You ought to nominate him for Air Force Special Operations Command (AFSOC)." In this conversation, John shamed the Chief and caused him to nominate me for AFSOC.

I was nominated to the AFSOC Commander and invited to meet with him for an interview. During my interview with the 3-star general, it became clear to me that I had encountered a man of faith. He shared with me his favorite passage of scripture as a Christian. It was a passage from the Old Testament, Isaiah 6:8, *"Then I heard the Lord asking, whom shall I send as a messenger to this people? Who will go for us? I said, here I am, send me."*

It was the most refreshing moment that I had experienced in many months. I knew that I was in the company of a man of great humility and faith. This too, was a God-thing.

I had a great visit with the general and was hired for the job as the Command Chaplain. I experienced more grace with the general during that short visit, than I had experienced with clergy, at the highest levels of the chaplaincy, in years! He exhibited fruits of the Spirit, spoken of in the New Testament, the book of Galatians 5:22, *"The Holy Spirit produces this kind of fruit in our lives: love, peace, patience, kindness, goodness, faithfulness, and self-control."* This was a "WOW" moment for me!

Inwardly, my spirit was rejuvenated, and I praised God for His faithfulness and marvelous grace! I left the US Southern Command in July 1998 after two good years of working with all branches of the military. Despite all the challenges, I left with wonderful memories!

Chapter 16

MY GREATEST CHALLENGES

U. S. Air Force Special Operations Command (AFSOC) Senior Years (1998–2002)

"Do nothing from selfish ambition or conceit, but in humility count others more significant than yourselves. Let each of you look not only to his own interest, but also to the interests of others."

PHILIPPIANS 2:3-4 (ESV)

While serving on my last assignment as the command chaplain at AFSOC in Florida, I received a call from a Chaplain Lieutenant Colonel who moved up to the Air Force Personnel Center (AFPC) or Chaplain Assignment Team. This team, led by a 06 chaplain, came to Washington, D.C., in October 1999 to meet with the Chief and Deputy Chief of Chaplains. During the meeting, there was a discussion about the position of Executive Officer to the Chief of Chaplains. Someone suggested an African American chaplain for the role. Chaplain Brigadier General Lorraine Potter rejected the idea, allegedly stating, "African American chaplains are good pastors and preachers but cannot do staff work." This remark was perceived as racist, leading to a call to me for investigation.

I called Chaplain Colonel John Blair and Chaplain Colonel Raymond Hart and briefed them on the allegation. We decided to confront the Chief of Chaplains at the upcoming annual Wing Chaplain Conference (11-15 October 1999) at Offutt AFB, Nebraska.

On October 11th, we had a 30-minute discussion with the Chief of Chaplains about the allegation. After the discussion, we felt it was clear that the Chief of Chaplains had no intention of dealing forthrightly with this issue. We then decided to take action by writing an official letter to the Chief of Chaplains as a matter of record to inform him of our position.

Another African American chaplain, who remains unnamed asked to see the letter sent to the Chief of Chaplains. He shared it with a contact in the Air Force, leading to a 4-star general's involvement. The General, upon seeing the letter, contacted the Chief of Chaplains and demanded an investigation into the allegations.

The Secretary of the Air Force (SAF) authorized the Inspector General of the Air Force to conduct the investigation of Chaplain Potter.

During the week of February 2000, I contacted the Chief of Chaplains and requested a meeting with him and the Deputy Chief of Chaplains. I indicated that Chaplains Hart, Blair, and I wanted to discuss some important African American Chaplain's issues with them. On 14 February, I received his response. The Chief of Chaplains indicated that our requested meeting would be on 20 March 2000. He requested us to forward any read-aheads or information we believed would be helpful in his preparation for the meeting. We made a conscious decision not to send anything, because we wanted to avoid their preparation of bogus answers, excuses, and irrelevant or inaccurate information for our very serious matters.

On 20 March we arrived at the office of the Chief of Chaplains and his staff, which included all the Division Chiefs. We sat down around the large conference table, and I proceeded to present our issues. We intentionally did not give them copies of the briefing until the end of the presentation. I requested their undivided attention, with no questions until the end of the presentation. At the end of my presentation, several people had questions and comments.

At one point, while Chaplain Blair was speaking, one of the colonels, a division chief, interrupted him. Chaplain Blair stopped making his remarks, pointedly looked at him, cleared his throat as he collected his thoughts and emphatically said, "Don't interrupt me when I am speaking! When I finish, then you may speak!" He did not mince words and never flinched! You could hear a pin fall on the floor! It was tense. I do not think anyone around the table would have dared to interrupt John Blair after that moment of drama. Quite frankly I was shaking in my boots! When the meeting was over, we made our way from the office. We had made our point, and we knew that they did not take too kindly to what they had heard! We emphatically stated that we had lost confidence in their ability to lead the chaplaincy!

Later we were told from a reliable source in that office that as soon as we left their office, one of the attendees asked for permission to contact the Judge Advocate General (JAG), whose office was in the same building. They wanted to see if they could bring insubordination charges against the three of us!

As word of our bravery in dealing with the highest level of the chaplaincy circulated, some of our peers labeled us as the three amigos. "There were others who likened us to the three Hebrew boys in the book of Daniel, 3:16-18, Shadrach, Meshach, and Abednego. As you may remember from your early childhood bible stories, these three were condemned, by King Nebuchadnezzar, to die in the fiery furnace. At some point our group became known as: Shadrach, Meshach, and a Bad-Negro. John was tagged as "a Bad Negro" because he was the most vocal, took no prisoners and would stand up with me to defend the honor and reputation of African Americans. He was a "BAD BROTHER!""

Just before the results of the investigation was scheduled to be released, I received a call from the head of the SAF/IG, a 3-star general, wanting to know if I could fly to Washington, DC, to meet with him regarding the findings of the investigation. This was March of 2000. I indicated that I would not be able to come until 27 April, when I would be in DC for a prior planned conference. I made sure that he understood that he would need to speak with Chaplains Blair, Hart, and me. I wanted him to know that we were a solid team! The plan was for him to meet with us in one of our billeting (similar to hotel) rooms.

He came to our room, and we were surprised when his Deputy, a 2-star general accompanied him. We were under the impression that we were meeting with him alone. He proceeded to share with us what the report of the investigation concluded. The report concluded that, "the allegations were unsubstantiated." I almost came unglued when he made the statement. I could not believe that they could arrive at this conclusion with all the information that had been provided. He knew he had hit a live nerve, to which he said. "Now chaplain, hold on, we know that something was said. Something happened in that meeting that was wrong, but we just could not prove it!" He further stated, tomorrow I will have the courier to drop off a copy of the report, then you will be able to see why the allegation was unsubstantiated.

The next day the report was delivered to me as promised. The three of us chaplains, went over the heavily redacted report and easily discovered what had happened for the allegation to be unsubstantiated. Because there were just a few people investigated, we could easily figure out who was speaking, even though it was heavily redacted as far as names and ranks were concerned. What we discovered was the Investigation Officer (IO), did a second visit with the African American chaplain and one other White chaplain. When the African American chaplain was questioned for the second time about the statement he heard from Chaplain Potter, he said, "Whatever I write down would probably be my spin on it, but I think, to the best of my knowledge that's as close as I can remember her saying." It must be noted that this was a change from his original statement in which he was very specific about what he had heard.

The Secretary of Air Force Inspector (SAF/IG) stated that we would see why the report was unsubstantiated. The above statement would register some doubt as to what he heard. When he mentions, "my spin," that becomes a killer statement. It basically implies uncertainty about what he heard. What are some possible reasons for his uncertainty? Would the fear of repercussions creep in? Would fear of not being promoted to 06 enter one's judgment? It could have been one of these or none of these. I will not sit in judgment of Chaplain _____, I was not there when it happened. So, I did what I believed I had a responsibility to do as a senior chaplain in the United States Air Force.

Following the investigation, I received a long e-mail from Chaplain Potter, dated 19 May 2000 with her being surprised regarding the statement she had made. I felt that in the email she simply tried to appease me. I also received notification of the completion of the investigation.

I then called for a reinvestigation by SAF/IG. That request was based upon a document from a Judge Advocate General (JAG) who had done an analysis of the Report of the Investigation.

We had an upcoming African American Chaplains Retreat to be held on the campus of Hampton University in Hampton, Virginia on 5 June 2000. The Chief of Chaplains planned to bring Chaplain Potter with him to give her the opportunity to clear her name. The three of us, 06 chaplains, did not believe it was in the best interest of our gathering, especially at that time to have Chaplain Potter present on the heels of the investigation. We believed that the investigation had been mishandled.

To this day, we have questions regarding the policies of investigation of senior officers. Does the Air Force Policy, regarding investigation of senior officers allow a colonel to investigate a 1-star and 2-star general? I had always heard that the Investigation Officer (IO) had to be equal rank to or higher than the officer being investigated. When I was investigated at Keesler by AETC/IG, I was a colonel. They did not send a lieutenant colonel to investigate me. They sent a full colonel equal to my rank to do the investigation.

Since my retirement, I have searched for a definitive answer to my question. In my search, if the investigation has named subjects, the Investigation Officer (IO) must be equal or senior in grade and rank to the most senior subject and not in the subject's chain of command."

We had always believed that decisions at the general officer level could be very political. We believed the Air Force would have received a great deal of blow back, had the charges been substantiated. If Air Force policy states that the Investigation Officer (IO) must be of equal rank or above the officer being investigated, then I submit that the Air Force has two systems of justice! Furthermore, if the SAF/IG knowingly made the decision to select the inappropriate grade of the IO for this case, then it seems to me that the Secretary Air Force/IG (SAF/IG) is part of the problem!

We believed that our retreat, designed for spiritual renewal, would be dampened with the presence of the Deputy Chief of Chaplains. I emailed the SAF/IG appealing the decision of the Chief of Chaplains. We were convinced that it would not make sense for Chaplain Potter to attend the retreat. It was not the proper time for her to visit. Something of this magnitude required a "less tense" environment. Despite our resistance, the top leadership rejected our request and decided to bring her to the retreat any way. The result of these exchanges between the Chief of Chaplains and me was "NO DEAL." The stars on their shoulders outranked the eagles on ours.

On the morning of 5 June 2000, Chaplains Hart, Blair, and I met with the Chief and Deputy Chief of Chaplains in Washington, DC. The purpose of the meeting was regarding Chaplain Potter's being cleared of the charges and able to get a 2nd star. A new initiative was presented. The initiative was to conduct a climate assessment of the Chaplain Service. The motivation for a climate assessment was directly related to the allegations against the Deputy Chief of Chaplains, Chaplain Potter, and the perception of her by African American Chaplains. The assumption was that perceptions were just the tip of the iceberg as related to racial discrimination in the Air Force. In other words, Air Force leadership felt so strongly about this issue that they wanted to assess how African Americans felt generally about this issue. The results of this assessment are astounding!

After this meeting we drove to Hampton, Virginia to the campus of Hampton University for The African American Chaplain's Retreat. It started with a social time, then moved on to the gathering of the African American chaplains. I was apprehensive because I was aware that the Chief of Chaplains was about to address the group about the issues concerning Chaplain Potter. All of us were polite and showed the Chief and Deputy Chief of Chaplains proper respect and protocol. After their presentation they went back to Washington, DC. None of us had any questions or comments. Their presentation was not for discussion, it was simply for sharing information. I am confident that none of our minds were changed by their visit or presentation.

The next day, during our retreat, we were privileged to hear from a few of the significant African American preachers and theologians from around the country. This was a very timely and inspirational opportunity because we needed a spiritual uplift. I had mixed feelings about the way things had happened and felt helpless to do anything about the situation. I was discouraged that we could not persuade the powers that be, not to attend at this time, but was glad to hear the preachers. I needed to hear words that lifted my heart and spirit.

During the retreat John contacted an African American, Baptist preacher from Boston, Massachusetts, who personally knew Senator Ted Kennedy. John had been made aware of the minister's connection to Senator Kennedy by a fellow chaplain who knew what had occurred in the situation with Chaplain Potter. John, Ray, and I met with the preacher, and briefed him on the allegation and investigation. After we briefed him, he assured us that when he got back to Boston, he would contact Senator Kennedy and request that the Senator not vote for Chaplain Potter's nomination of a 2nd star. We knew that if there was one opposition to confirming her, she would fail to be awarded that 2nd star. I suppose it just was not to be. Before the preacher had the opportunity to contact Senator Kennedy, the preacher had a massive heart attack and died.

Results of the Study confirmed our beliefs about racism in the Air Force. The report on the Study showed that 97% of African American Chaplains sensed or directly experienced racial discrimination. When asked whether they perceived that a "Good Old Boy System" existed in US Air Force Chaplain Service, 91% acknowledged that they perceived it as existing and contributing to assignments, career progression, and promotion opportunities.

Of particular interest is the fact that some of the respondents had low expectations that any substantive changes would result from the climate assessment. The summary, at the end of the assessment, included some positive comments. Unfortunately, most of the comments were dismal or pessimistic. This is an example of what I mean, "Do I think the results of this survey will cause anything to change within the AF Chaplain Service? Absolutely not!"

As I write this book, 23 years later, the "promised fixes" as indicated by the Air Force Chief of Staff, never fully occurred. In my opinion they gave African American chaplains a "worthless check" that bounced. Further, my sources of African American chaplains currently on active duty indicate that conditions for assignments, career progression and promotion opportunities have deteriorated and are worse today than they were when I retired in 2002. Additionally, no United States Air Force African American chaplain on active duty has broken through the glass ceiling to reach the rank of general officer to this date.

When the Chief of Staff analyzed data from the survey, he saw the disparity and plight of African American chaplains and others, on a lesser scale. He directed the Chief of Chaplains and the Deputy Chief of Staff for Personnel to examine the issues highlighted in the survey and develop an action plan. The plans were to be completed by 1 Oct. 2001.

After the climate assessment was completed and the results were so revealing and catastrophic, it was determined that a leadership diversity task force would need to be created. The Chief and Deputy Chief of Chaplains had the responsibility of chairing the task force. The task force was composed of commands and several handpicked chaplains from each command as representatives. Each command was to include African American chaplains to represent the command. Some small commands did not have African American chaplains available, such as AFSOC, my command. We attempted to appoint African American chaplains from other commands to ensure proper representation on the task force. I was met with push back when I selected an African American from another command to represent AFSOC. He traveled to the meeting place, but was not allowed to be a part of the diversity task force.

We went to several task force meetings in which proposals were made to rectify some of the prevailing issues. At the end of the second meeting, we were asked to provide feedback and evaluation of the task force proceedings. However, there were always questions as to the integrity of the leadership in following through on the proposals.

There are documents from White chaplains, including a command chaplain, who shared serious concerns about the behaviors of the Deputy Chief of Chaplains. From all indications, since that time, nothing of any substance has changed. After more than twenty years, African American chaplains get to a certain point in their careers, or are assigned positions in which there is no upward mobility, and the glass ceiling remains shatter-proof for higher ranks.

The Deputy Chief of Chaplains was cleared of charges against her and later received her 2nd star and became the Chief of Chaplains.

Chapter 17

MY MILITARY CAREER ENDS

*"I have fought the good fight, I have finished the race,
I have kept the faith."*

Тimothy 4:7 (ESV)

In the late summer of 2001, I had begun preparations for my retirement. I wanted to have a celebration with a retirement dinner. I had to decide who to ask to do my retirement ceremony. I debated it over and over with my good friend, John Blair. He highly recommended that I invite General _____, who had interviewed me for the position of command chaplain at AFSOC. He hired me, however, by the time I arrived at Hurlburt Field, he had left to become the Commander In Chief (CINC) at Central Command in Florida, where he had been promoted to a 4-star general. I had deep reservations about inviting him since I had not worked for him one day. There is no doubt in my mind that John Blair, behind the scenes, had made a call on my behalf to the general, since they were good friends. I reluctantly called him and requested him to consider coming back and presiding over my retirement service. He graciously agreed to come and do it.

September 11, 2001, as you might well recall, was a day of infamy. I sat in my office that morning watching the news when the planes flew into the towers in New York City, as well as the Pentagon in Washington DC. I sat shocked, in disbelief! Days later, I realized that Central Command had the responsibility of prosecuting the war, brought on by the actors of 9/11.

While my retirement was mandatory because of my age, there were others on the base, some of whom had requested retirement, approved, but were rejected due to 9/11. Some had already had their household goods picked up, but now were in the process of being deployed! There was no question that I may need to go back to the drawing board to look for someone else to retire me. I was so relieved that General _____, with all that he had on his plate, still stated that he would be there to do the honors.

The time came, when I was within 30–45 days of official retirement, the inmates were still trying to take me down! I was sitting in my office when the phone rang. It was the secretary of the 3-star general, my boss. She said to me, "Chaplain Beamon, General _____ wants you to report to him immediately!" I was shocked because I had never received a call like that, never! I could not begin to imagine what I had done for such a call. Nothing came to my mind as I took inventory of any actions or involvements, I had with anyone. I made my way to the general's office. The 1-star Deputy Commander escorted me into the 3-star's office, where I sat at his large table in his conference room. I was there alone for about five minutes, when both generals came in and sat at the table. From his facial expression I could tell he was upset. He was "tight-jawed and tense. He proceeded to verbally reprimand me. He was upset about a letter I had written to the Deputy Chief of

Chaplains. This letter had been written and e-mailed to the Deputy Chief more than a month ago, with copies sent to the Chief of Chaplains and the 3-star, my boss. In the letter to the Deputy Chief of Chaplain I had criticized him for his unprofessional behavior as a senior officer at the Diversity Task Force meeting as well as at the African American chaplains' retreat. I did not mince words and I had no problems in making sure that top leadership knew it.

When the 3-star finished, I addressed him in this way, "Sir, with all due respect, I admit that what I said in my e-mail to Chaplain _____ could have been stated in a different way, however sir, what I said was true. I will make this commitment to you sir; from this day forward, I will have nothing to say, no communication with Chaplain _____ I give you my word."

This was the statement he needed to hear from me. That statement assured him that I would not cause any more trouble before I retired. "Good Trouble?" He stated that he had a letter of reprimand in his desk drawer that he had intentions of issuing to me for charges of insubordination of a senior officer, who happened to be his friend. However, with my explanation he would not issue it. For the life of me, I tried to figure out what was going on. The letter had been received by the Chief of Chaplain, the Deputy Chief of Chaplain, and the 3-star, my boss, a month ago. So why, suddenly, did everybody focus on it 45 days before I was due to retire? I could understand their concerns if they had responded in this way a few days after receiving the letter. My feeling was that they wanted to "mess up" my retirement by having me removed before my retirement services. They had plenty of time before then, to raise any concerns. Having been in the military for 27 years, I knew better than to blindside my boss. That is why I sent him a courtesy copy of the letter.

The other thing of which I was keenly aware had to do with the relationship between the Deputy Chief and my boss. I had been told by my boss in our initial meeting that the two of them were long time friends and classmates at the Air Force Academy and they were in pilot training together. I knew they were friends, but my professionalism dictated that I do the right thing. I felt it was necessary for him to know how his friend's unprofessional behavior, explosive temper, impacted other chaplains.

I had just gotten back to my office when my phone rang again. It was my boss's secretary again. She told me that the general needed to see me again. I made my way back to his office. He asked, "Walt, do you know of any reason why your successor, in Germany, needs to report 30 days early to relieve you?" I indicated that I absolutely knew of no reason. I stated that I had everything in place for his arrival and I could not think of any reason his service was needed now. He then said, "Okay chaplain." That was all he needed. By the time I got back to my office, my phone rang again. This time it was my successor calling from Germany, wanting to know what was going on at AFSOC. I responded as if I did not know. He said, "I just received a call from the Chief of Chaplains, Potter, saying that I needed to report to AFSOC in 30 days! I'm not ready to come that early and I'm just wondering what the heck is going on there!"

The bottom line, the Chief of Chaplains was attempting to have me fired and relieved of duty 45 days before my official retirement! Thank God that the 3-star general, my boss, would not agree with her plan. They meant it for evil, but God meant it for good.

On the evening of my retirement dinner, there were approximately 150 people in attendance. Numerous chaplains, with their spouses, from across the country both African American and White, along with a large number of our church family, First United Methodist Church, Crestview, Florida. The bell choir from the church provided beautiful patriotic music for the event. As the activities for the evening proceeded, I reflected on the challenges of my military career and the beautiful event that night. I specifically mentioned that during my retirement, one of my goals was to write a book, chronicling some of my experiences during my 27 years in the Air Force chaplaincy.

On Monday morning 10 December 2001, the crowning event of my retirement occurred. Thirty minutes before the ceremony began, General _____ graciously met with my immediate and extended family members. He made all of us feel so much at ease as he visited with us and learned some details of our lives. Later as he presided over the retirement ceremony, I sat amazed as he personalized the moment by remembering family members by name and spoke about my early years and career in glowing accolades for 15 minutes without a single note!

On the front pew of the chapel, across the aisle from my family, sat my boss and his friend, the Deputy Chief of Chaplains. It is customary for a formal invitation from a chaplain colonel, be extended to the office of the Chief of Chaplains. I did not extend such an invitation. I had lost respect for senior leadership and would not display hypocrisy by extending an invitation.

As the general spoke with such eloquence, I would periodically glance over toward my boss and the Deputy Chief of Chaplains. It was obvious from the looks on their faces that they were most uncomfortable with what they were hearing from the 4-star general. I could not in my wildest imagination visualized the beauty and grandeur that day would bring. It was a wonderful way of celebrating the end of 27 marvelous years of service to God and Country! "To God be the glory for the great things He has done!" Would I do it all over again? YOU BETCHA I WOULD, HANDS DOWN! Despite the pitfalls, the ups and down, the tough calls, I received comfort from the Word of God. The Apostle Paul asks this question, *"Who shall be able to separate us from the love of God in Christ Jesus? Shall tribulation or distress, or persecution or famine or nakedness, or danger or sword? No, in all these things we are more than conquerors through Him who loved us. Nothing will be able to separate us from the love of God in Christ Jesus."* Romans 8:35 :37 (ESV)

I live my life with only one regret, looking back over those years, I have come to realize that I needed a closer relationship with Jesus Christ during my journey. I wish to end this chapter of my life with some powerful words from President Lincoln that he spoke at the end of the Civil War, to paraphrase, "with malice toward none, with charity for all." Looking back over my journey, I can truly say, "The Inmates In Charge" made me a better person! God's intervention in the ending of my military career with that beautiful retirement event purged me of bitterness. Comments, and words of encouragement from those in attendance filled my heart with joy. I could not, in my wildest imagination, have experienced a better send-off from the military. THANKS BE TO GOD!

RETIREMENT CELEBRATION

Chaplain Colonel Raymond Hart (Deceased)

Chaplain Colonel John Blair (Deceased)

Retired Master Sergeant Sadie Rebecca May Loving-Jackson

Chapter 18

BACK TO THE CHURCH

"Let the peace of Christ rule in your hearts, since as members of one body you were called to peace. And be thankful."

COLOSSIANS 3:15 (ESV)

For many years, prior to my retirement, I heard the expression, There is life after the Air Force." Truer words have never been spoken. While trying to unwind and prepare myself for civilian life, Ikie said to me more than once, "Walt you are going to have a difficult time adjusting to civilian life! I know that you have been accustomed to being in control, giving orders and having people respond to you in a commander's role, but that is coming to an end soon!" She further pointed out, that I was getting ready for the "real world." Boy, did she know what she was talking about. People did not call when they promised or come when they made appointments. If I indicated my disapproval, their response was indifferent. This was now foreign to me because, generally speaking, I had become used to people honoring their commitments. All those many years, she had worked in the civilian world and knew very well what it was like.

Back in 1998, when we built our retirement home in Crestview, Florida, there was a sub-contractor finishing work on the house. He was Dan Bradford, a local White fellow, and very friendly man. We struck up a conversation one day, when out of curiosity, he asked me if we were members of one of the local churches in Crestview. I indicated that we were brand new to the area and had not affiliated with any church at that time. He immediately invited us to his church, First United Methodist Church. His invitation was not a foreign idea, because we had been members of Perine Peters United Methodist Church back in Miami, Florida. I indicated that we would be looking for a church with a strong youth program, because we were raising our two grandsons, Monty and Devan, 8 and 6 years old, respectively. Dan, spoke empathically about their youth program. A few weeks later, after settling into our new home, we decided to visit the church. I knew it would be a predominately White congregation, but so was Perine Peters.

When we walked into First United Methodist, we discovered that we were the ONLY African Americans in worship. We sat near the back and could not help but wonder if we had made the best choice. Everyone noticed us. However, during the Passing of the Peace, a form of fellowship in which people exchange greetings; we were all acknowledged in a very friendly manner. Young and older members welcomed us. During the Childrens' moment with the pastor, Wesley Spivey, gave an invitation to all the children to comedown front. Our grandsons responded and went up with the other children. After the children's moment, they were dismissed to attend children's church. Our boys went also. I was a bit surprised when Pastor Spivey gave his sermon. He was energetic in his delivery and the sermon was really good. I had expected a sedate, dry sermon. Wow, what a pleasant surprise!

At the end of the service, parishioners came in small groups to welcome us and invited us back for future services. Ikie and I felt genuine warmth from them. On our way home, our boys were really excited about their experiences in children's church. They said to us, "Grandma, grandpa, we love this church!" So, we started attending regularly. After about six months, I joined the Sanctuary Choir. Ikie joined the United Methodist Women's organization and had a great experience sharing in their various activities. She eventually was elected as president of one of their "Circles."

After I retired from the Air Force in 2001, Dr. Richard (Dick) Wright, the newly appointed pastor, approached me about becoming his associate pastor. The previous associate had left for active duty as a chaplain in the Air Force in the middle of the conference year. Dick had some medical issues and needed help. Dick and his wife, Linda, knew that I had retired from the Air Force. We had conversations at church events, so they also learned from these conversations that I had pastored several churches prior to entering the military. Linda suggested that Dick ask me if I would be interested in taking on the position of associate pastor. He was hesitant because he was not sure I wanted employment, having just retired from the Air Force, but he did.

Honestly, I was surprised he asked me. I had some serious reservations. I considered the demographics of the church. It had been an established congregation for many years. They had "us" as members and eventually a few other African Americans joined the church. In its history, we were the first African American members. I had questions about being an African American associate pastor. I recognized that many of the parishioners were quite conservative. Another reservation I had was the fact that I was not a member of the United Methodist

Conference. Functioning in a predominately White congregation was not intimidating to me. Throughout the many years of service in the military, 75% if not more of the people I interacted with were White. The military was simply a microcosm of the larger civilian society.

My greatest concern was whether I would be accepted and respected in the role of associate pastor by the membership. When we met to discuss my decision, I asked him, "Dick, are you sure about me working here?" He understood the implications of this question and assured me that he was sure it would be fine. I told him I would accept the position with two stipulations. First, you must understand that I am retired and there will be times when I simply will not be available. I like to golf and plan to do more golfing in my retirement. Second, be aware that when you retire, in two years, I will step down from this position. He was okay with both provisions. So, after much prayer and thought, I decided to accept his proposal as associate pastor on a part-time basis.

Before we could seal the deal, He had to bring the proposal for my employment to the church administrative board for approval. I clearly remember the Sunday morning he met with the board. Despite his assurances to me about my acceptance, he was not completely sure that there would be enough positive votes to pull it off. But after the presentation of the proposal and vote, he was filled with excitement. The board thought well of his plan. They voted unanimously to bring me onto the staff.

My job description included typical duties. I had the opportunity to preach in two of the three worship services one to two times a month. We alternated leading the service or preaching for worship services. I also assisted with visitation of sick and shut-ins, and conducted weddings, funerals, and baptized babies. I became aware of an increase in church attendance when I came on board. I was never quite sure why the attendance seemed to be more on the Sundays I preached.

Perhaps it was out of curiosity or maybe it was due the fact that my style of preaching was different from most White preachers. I always tried to do my very best in the pulpit regardless of who happened to be in the pews.

I spent a great deal of time preparing sermons and prayers. I felt I owed it to my Heavenly Father to preach the unsearchable riches of Jesus Christ. I was reminiscent of the text of my first sermon, back in Chicago on 25 August 1966 from the gospel of John 15:16, *"You did not choose me, but I chose you. I appointed you to go and produce fruit and that your fruit should remain, so that whatever you ask the Father in my name, He will give you."* (CSB) Each time I stood behind the sacred desk, I longed for an anointing of the Holy Spirit to rightly divide the word of truth.

During my years of retirement, as I alluded to in the introduction, a group of African American chaplains came together in Crestview to organize a nonprofit organization, United Military Chaplains Association. As a retired senior officer, I was one of the founding members. Serving in this capacity required significant writing and teleconferencing with other senior retired chaplains.

When Dr. Wright retired two years later, I let it be known to the official board that I too would be stepping down. This had been the agreement when I took the job. Our rationale was that the incoming senior pastor should have the freedom to choose whom he would prefer as his associate. When I made this known, an officer of the board said, "Oh no, you can't go now, we need some continuity." They pleaded with me to reconsider and give them one more year. I agreed to do so, however, as the scripture reminds us, *"Now there arose a new king over Egypt, who did not know Joseph."* (Exodus 1:8 ESV)

A new senior pastor succeeded Dick, from Citronelle, Alabama, Pastor Bruce Scheffield (Deceased). He was uncomfortable with an African American serving as his Associate. He had been there for weeks without a conversation about me working with him as the Associate. Finally, I confronted him in a non-threatening way about how I was feeling with this lack of communication and clarity concerning his expectation of me and my role.

After a few months we developed a workable relationship. The relationship was quite different from the one I shared with Dick. We did not socialize on any level or make small talk or make plans as I had done with Dick. I did not enjoy the work environment as I had before.

I made the decision to step down after a year. He wanted me to make the announcement to the congregation instead of him. He also wanted me to make sure, in making the announcement, that I made it clear that the decision was strictly my idea and not his.

One of the highlights of our membership and most memorable occasion came on 13 August 2013. This was the celebration of our Golden Wedding Anniversary (50 years of marriage). We were blessed to have about 250 people attend from across the country. Our two adult children and three of our five grandchildren were also present. It was held in Ft. Walton Beach at the beautiful Sound Side Center. It was a day to remember. After more than 60 years together, I am so thankful that I endured the heat back in 1963.

THE WHITE HOUSE
WASHINGTON

Congratulations on your 50th wedding anniversary! An enduring marriage like yours is truly golden. Your support for each other through the joys and challenges of your half century together is an inspiration to us all.

We send our best wishes to you on this wonderful occasion.

Sincerely,

In June 2014, the congregation honored me by awarding me with the title, "Pastor Emeritus." We were members there for 16 years. They became like family to us; we knew their children and grands and they knew ours. One family said they were adopting us. This particular family always invited us to attend their Christmas Eve family time in their home. They always had delicious food. We would gather there, then go to church for the candlelight worship service. If we were in town, they expected us to be with their large family for their celebration. They were very sensitive to us not having family in the area at times like Christmas. We have wonderful memories of this congregation just as we have of our first small churches in Virginia. We still have good relationships with many of them, communicating by e-mail, cards, and phone calls.

Chapter 19

MY SPIRITUAL
JUSTIFICATION

"Therefore, since we have been justified by faith,
we have peace with God through our Lord Jesus Christ."
ROMANS 5:1 (ESV)

After twenty plus years, I have had a great deal of time to reflect and think about my journey. Perhaps the most prominent realization that comes to mind is the knowledge that this book could not have been written by me early in my retirement years.

My mindset was not such that I could have produced a book with a meaningful purpose. I now know that it would have projected anger and vindictiveness would have blanketed the script. I was not ready to write anything of any substance. After my retirement, as mentioned earlier, I became one of the founding members of our non-profit organization, United Military Chaplains Association (UMCA). I was determined to find some way to influence those in leadership positions within the chaplaincy, to do the right thing toward all of its chaplains, but especially African American Chaplains. I would be up into the wee hours of the morning, thinking, trying to develop strategies, trying to figure out ways that we might make a difference for young brothers and sisters in the chaplaincy. This was an institution that I loved being a part of for most of 27 years. I was passionate. My mind was preoccupied with thoughts of what might work day and night. I can remember throughout my career, as I went through rough times, somehow the good Lord would take my mind back to Chicago, a place that stood out as a spiritual marker for me. Chicago was the place where I first saw the light, where I had a lasting encounter with Jesus Christ! It was in Chicago that the gospel of John 16:33 took on a special meaning in my life. It states, *"I have said these things to you, that in me you may have peace. In this world you will have trouble. But take heart! I have overcome the world."*

Repeatedly, I sensed God's assurance that He had called me to this ministry. I knew that it was His doing, not mine. He had called me and said that I should go out and bear fruit! With each step along the way I sensed the nearness of His promise. I sensed it as I was led to Richmond, Virginia, Atlanta, Georgia, the Air Force Chaplaincy, Crestview, Florida, and everywhere I have been. I have borne fruit wherever God has placed me.

About seven years ago, I received a call from a member of the second little church I was assigned to as a student minister, during my second year in Seminary, St. Paul AME Zion Church, Meredithville, Virginia. She was a young teenager at the time we were there. She had spent years searching for us. She wanted to let both Ikie and me know just how much our ministry had impacted her during the many years since we were there.

When we moved to Atlanta a few years later (1970), a young family who had been in the US Army, moved to Atlanta for retirement. They joined our church, Shaw Temple AME Zion Church. Charles and

Marian Tunstall and their children were a great blessing to the church and to us. Their oldest son, Stanley was a young fellow at the time. Fast forward from 1970 to 2009 a period of 39 years. I received a long letter from Stanley (Stan), which is a keepsake. He has granted me permission to share excerpts from his letter, which I want to share here. It is dated 12 July 2009. During those 39 years he had grown, matured, a man of stature and now was ready for retirement after spending 29 years in the US Army, having reached the rank of colonel.

Dear Pastor Beamon,

I don't know if you remember me, but I am the oldest son of Charles and Marian Tunstall. I was one of many boys running up and down the aisles of Shaw Temple when the church was still on Hightower Road.

I don't think I've seen you since 1981, the year you wrote a letter of recommendation for me to attend Officer Candidate School.

I was blessed to find my passion and calling as a soldier. As I was thinking about my career, it occurred to me that much of my success was due to your support and inspiration. You gave me the spiritual foundation needed to overcome the challenges and difficulties associated with a military career: multiple separations from family, 14 change of station moves, and four overseas tours.

As I get older, I often reflect on my early years at Shaw Temple: the Sunday School classes, the gospel music, Vacation Bible School, and your passionate sermons. As a young boy, there were many Sundays when I played with my church money or made paper airplanes, but somehow your messages touched me and the seeds of my faith in Christ were planted. With this perspective, I left home with a basic understanding of who I am in Christ, why I am here, where I am going, and how I should treat others. Some of the most profound and enduring lessons in my life are the ones that came out of your pulpit. In fact, I consider you one of the greatest teachers I ever had. It has taken many years, but the fog is lifting, and I can see God's hand across the span of my life, slowly leading me to the ultimate redemption."

Stanley had much more to say about his wife and children and how they are such a blessing to his life. I appreciate knowing my impact on these two individual's lives.

They are two of many who have contacted me over the years about how their lives were blessed through our ministry. I've had others who have contacted me asking if I recalled a certain sermon I preached at a particular chapel, and they were saved as a result of the message. The Apostle Paul, 1st. Corinthians 3:6 says, *"I planted the seed in your hearts, and Apollos watered it, but it was God who made it grow."* (NLT)

Despite the many challenges, pitfalls, obstructions and injustices, African American chaplains still achieved and gave great service. We fought valiantly, stood tall and accomplished much. "To conclude this chapter, I draw inspiration from the renowned poet Maya Angelou and her celebrated work, 'Still I Rise.' Although I cannot include the poem itself, the spirit of her words echoes throughout this chapter."

Chapter 20

FINAL THOUGHTS — IMPACT AND LEGACY

"If any of you lacks wisdom, you should ask God, who gives generously without finding fault, and it will be given to you"

JAMES 1:5 (NIV)

I want to conclude my memoir by talking about something that I believe is the hallmark of everything that I've tried to highlight as part and parcel of my life. It is the subject of diversity, which is defined as the practice or quality of including or involving people from a range of different social and ethnic backgrounds and of different genders, sexual orientations, etc.

During the years that I served in the US Air Force, African Americans represented 22% of the total force. African American chaplains were 8% of the chaplain corps. The sheer numbers automatically meant that there were challenges of inclusion from day one. Getting your slice of the proverbial pie was always going to be an uphill journey. I would suggest that the same challenges faced in civilian life were front and center in the military community. When policies or laws were enforced, usually military powers tended to enforce laws on the book with more diligence than in the civilian community. Racism and discrimination in the Air Force seemed to have plagued African American chaplains as well as those in other careers. It tended to be more covert than overt and blatant. Although policies written on paper stated zero tolerance for discrimination, the enforcement of the policy was determined by the commander and his or her advisors in charge. In my opinion, the inclusion or exclusion of African American chaplains by senior leadership in the chaplaincy depended on their experiences with race relations from their childhood years. It only seems natural that whatever the person in charge had experienced or was exposed to when growing up, often determined how they enforced written policy.

Diversity in the Air Force chaplaincy was limited at best. Yet, I have learned that there are some great benefits to diversity. A diverse organization sparks innovation. In other words, people feel free to think outside the box without fear of repercussions.

- A diverse organization is likely to come up with better answers for a given problem, compared with an organization where diversity does not exist.
- A diverse organization improves team performance. The whole concept of teamwork takes on a new meaning. There is less chance of bickering and in-fighting within that organization when teamwork is encouraged and practiced by those leading the organization.
- A diverse organization tends to make better decisions because people feel their opinions matter, they are appreciated and valued.

- A diverse organization in which equality is intentional leads to a stronger organization because each person is committed to the welfare of the other.

If, in fact, African American chaplains make up 8% of the total chaplain corps compared to 92% of Whites and others, then it is a stretch to believe that diversity is a reality within the organization. Let's face reality! The real challenge boils down to the concept of power! Those who have it are not willing to share it! The glass ceiling will remain a glass ceiling, reinforced with steel if senior leadership remains uncommitted to diversity. To have a healthy chaplain corps the concept of diversity as I mentioned, must be intentional, starting at the top, senior leadership—that is committed not only by word but with "deeds." In other words, when senior leadership is committed to the concept of diversity in its fullest sense, then and only then, can the chaplaincy become a healthy, thriving organization. When diversity is lacking or severely limited, individuals within the organization feel a lack of connection to others in the organization. Morale will seldom if ever be high in such an organization. I know a number of African American chaplains who retired with bitterness, anger, and resentment, because they never really felt connected and rewarded by the organization, the organization which was supposedly the "moral arm" of the Air Force.

Throughout my career, I attempted to establish and maintain healthy and wholesome relationships with both African American and White chaplains. My spiritual values dictated inclusion. However, I was not naive in thinking that all my attempts were reciprocated. When the issue of relationships with White chaplains surfaced, especially my counter parts or those I was competing against, I had to be very careful and perceptive in determining whom I could trust. It was human nature that I could only go so far in relationships due to the "trust factor" and limited diversity. It was important to be sociable and friendly, however, I always remembered the challenges of past generations and the trust factor.

This reminds me that at the end of my first assignment at Keesler, I learned that I had to be observant and shrewd in my relationships. This reality applied whether I was dealing with my counterparts or the line officer in another career field. The same principle applied whether I was dealing with my supervisor or boss or the Wing Commander. The bottom line for me was to work hard, be professional, treat my counterparts with respect, but I always tried to stand up for what I believed was just and right to the best of my ability!

When I was motivated and led to enter the Air Force chaplaincy, I had no grandiose ideas about rank. I did not come into the chaplaincy with the goal of becoming a colonel. My goal was to become the best and most effective chaplain I could become. I never sought after or kissed up to any commander or anyone in leadership position, never!

I will never forget a conversation I had with a commander. He asked me, "Walt are you staying out of trouble?" I thought he was joking at the time. I responded, "Yes sir, at least I'm trying!" To which he replied, "Well, you're not doing your job!" I never forgot those words. After that conversation, I couldn't help but ponder about the impact and truth of those words. At that point I was a major in rank. Those words from that commander changed the trajectory of my focus and simmered in my subconscious for years. When I reached the rank of a senior officer, I could hear those words loud and clear. I remember Chaplain Colonel I V Tolbert (deceased) once making the statement to me, "Walt, when you become a senior officer, you need to TAKE CHARGE! If you don't take charge, someone else will!" Another good friend, Chaplain Colonel (Ret) Wilfred Bristol, often reminds me that there is a significant difference between a colonel and a 06. Both wear eagles on their shoulders but there is a huge difference in the impact that a colonel makes over a 06.

Hopefully, I grew to become a colonel by the grace of God. It seems as though I was always making trouble, especially the last seven years of my journey in the Air Force. I would cautiously borrow a phrase from the late, great Civil Rights activist and Congressman, John Lewis. That phrase is "good trouble." I in no way would compare myself with such a distinguished servant of the people! In retrospect, I will say that I responded to anything I saw as a senior officer which appeared to be bogus and would not pass the smell test. I suppose that catapulted me into the class, some would refer to as a troublemaker! There is no doubt in my mind that the inmates in charge saw me as a troublemaker and kept me in their sight.

Another thing that I learned along the way, was the fact that I had to learn to choose my battles. There's an old adage that I recall hearing, that went something like this, "You cannot afford to fall on your sword with every battle you face, otherwise you'll bleed to death!" I learned to choose my battles with caution, thereby falling on my sword only when I was convinced it had lasting value and worth the cost.

When I entered military service, more specifically, service to God and Country, the learning curve was rather steep for me because I had no military background. I found it very difficult to be a part of an organization and not know the proper questions to ask due to my limited information. It's especially difficult if you are a member of an organization and your sponsor limits the information he shares with you. During the early days of my Air Force career, we did not have mentors assigned, or someone to show us the way and give information pertinent for growth and development.

As I conclude this chapter, there are a number of areas I wish to highlight in terms of lessons learned, which hopefully would be of value to chaplains in future generations.

- Be proactive in demonstrating a willingness to hear constructive criticism and feedback in order to move forward.
- Request help or assistance when needed and do more listening than talking.
- Never blindside your boss by taking action without their authorization.
- If you're not sure about a situation you are facing, seek out someone you trust, someone you deem knowledgeable and ask questions and more questions.
- Use the chain of command and don't become a "hot shot," creating problems for yourself! Steer clear of being known as a know-it-all person. You will not make friends if you fall into that trap.
- Avoid the trap of being known as one who operates with hidden agendas. Try to make transparency part of your modus operandi and you will likely be perceived as a straight shooter. Transparency is very important and necessary in garnering respect and integrity.
- Never bring unnecessary attention to yourself.
- Respect everyone, but never be known or perceived as a doormat for others to walk over. You'll never be disappointed when you look at the man or woman in the mirror.

RECOMMENDATIONS TO SENIOR LEADERSHIP

As this book has outlined various challenges faced by African American chaplains in the Air Force Chaplaincy, it is imperative to propose sustainable and positive change recommendations.

These recommendations result from collaborative efforts among retired Command chaplains, Assignment Team chaplains and other senior chaplains. If these recommendations are taken seriously and implemented, they will initiate a transformative process for a more equitable and effective Chaplain Corps.

Historical Context and Current Realities:

The Air Force Chaplaincy has been plagued by systemic issues, including "White Privilege" and incompetence at senior levels, leading to a weakened institution. For years, there has been a lack of effective leadership, and it is our view that the Secretary of the Air Force must ensure accountability within the Chaplain Corps.

Notably, since the United States Air Force became a separate department in 1947, African American chaplains have yet to be promoted or even seriously considered to fill the positions of Deputy Chief of Chaplains, 1-star general, Chief of Chaplains, 2-star general, which is a stark contrast to the Army and Navy, which have both promoted African Americans to these top positions.

In addition, there was an allegation that the Deputy Chief of Chaplains insulted a group of African American chaplains during an official public forum. These charges were substan-tiated, resulting in the Deputy Chief of Chaplains being relieved of duty. Unfortunately, this is not an isolated incident; similar situations have occurred at least four times. Despite the dismissal or forced retirement of these Deputy Chiefs of Chaplains, qualified and reputable African American chap-lains have consistently been overlooked for these positions.

Another significant issue I would like to highlight concerns the unusual promotion of a Colonel to the position of Chief of Chaplains, a 2-star rank, bypassing the traditionally intermediate role of Deputy Chief of Chaplains, a 1-star position. This deviation from standard practice is unprecedented in the Chaplaincy. Once again, this bypass resulted in the exclusion of many competent and proficient African American chaplains from consideration for these high-ranking positions.

I sincerely believe that had there been a level playing field and African Americans had been given an opportunity to ascend to the position of Deputy Chief of Chaplains, we would have made it to the top! We would have become the Chief of Chaplains. The mismanagement of the Chaplains Corps has been gross.

Recommendations for Change.

1. **External Management:** Temporarily place the Air Force Chaplaincy under the management of a different department, separate from the Chief of Staff and his subordinates, until confidence in its leadership is restored.

2. **Inspector General's Review:** A review should be mandated by the Inspector General every two years to monitor the promotion rates, assignments, and Professional Military Education (PME) selections of minority chaplains, which is crucial for career advancement.

3. **Leadership Development:** Identify and nurture three to four African American chaplains at the Major (0-4) level who demonstrate the potential for senior leadership. Provide them with a broad range of

responsibilities and assignments that match their abilities.

4. **Diversity and Inclusion Training:** Implement a mandatory Diversity and Inclusion course at each of the USAF Chaplain Corps College level to foster an understanding of diversity and sensitivity.

Conclusion:

This work represents a heartfelt endeavor to address longstanding issues within the Air Force Chaplaincy. With the completion of this book, I am filled with hope and faith that meaningful change is possible. While the future is uncertain, I am confident we can achieve a more inclusive and effective Chaplain Corps with guidance and commitment.

ACKNOWLEDGMENTS

I wish to acknowledge the following people who helped make this book possible. First, by far and foremost, my wife and soulmate, Ikie. She spent many days and nights helping me bring clarity to issues, circumstances, and experiences, which I knew in-depth but needed her help to make readable.

Second, to Chaplain Colonel (Ret.) Wilfred Bristol, who was always ready and willing to assist me when needed. He, too, had experienced much of what I did on active duty in terms of racism and injustice. I could always depend upon him for advice as I wrote this book.

Third, I must mention several other African American chaplains, including, but not limited to, Chaplain Lt Col. (Ret.) Bennie Liggins, Chaplain Lt Col. (Ret.) Shelby Taylor, Chaplain Lt Col (Ret.) Larry Keith and Chaplain Lt Col (Ret.) Theodore (Ted) Henderson.

Fourth, I extend our heartfelt gratitude to Mrs. Subrena Alford. Your invaluable contribution made this book a reality, and I am immensely grateful. To God be the glory!

ABOUT THE AUTHOR

Chaplain Colonel (Retired) Walter E. Beamon's 27-year journey as a US Air Force chaplain and trailblazer spans assignments across the globe. He became the 9th African American Chaplain (USAF-active duty) to achieve the rank of colonel. He achieved a number of firsts during his service in the military and civilian life.

- First African American Chaplain to serve as Wing Chaplain at Kessler AFB 1993 to 1997
- First African American Chaplain to serve as Command Chaplain HQ USSCOM 1997 to 1999
- First African American Chaplain to serve as Command Chaplain HQ AFSOC 1999 to 2002
- First African American minister to serve as Associate pastor and Pastor Emeritus, First United Methodist Church in Crestview, Florida.

He was the 9th African American Chaplain (USAF-active duty) to reach the grade of colonel in the 77 year history of the USAF. Only 28 have reached this milestone and none higher than a colonel to date.

Upon retirement from the USAF in 2002, he accepted a position as Associate pastor at First United Methodist Church in Crestview, Florida, a predominately White, very conservative congregation from 2002 to 2005. In 2014 the congregation honored him with the title of Pastor Emeritus. Chaplain Colonel Beamon is a devoted family man residing in Madison, Mississippi, with his wife, Ikie, and his legacy extends to two children, five grandchildren, and three great-grandchildren. His dedication to faith, family, and service remains unwavering.